T0298531

Computational Intelligence for Information Retrieval

Sensors Communication for Urban Intelligence

Series Editors: Mohamed Elhoseny and Xiaohui Yuan

Biometric Data in Smart Cities
Methods and Models of Collective Behavior
Stepan Bilan, Mykola Bilan, Ruslan Motornyuk and Serhii Yuzhakov

Computational Intelligence for Information Retrieval
Dharmender Saini, Gopal Chaudhary, and Vedika Gupta

For more information about this series, please visit: https://www.routledge.com/Sensors-Communication-for-Urban-Intelligence/book-series/SENCOMM

Computational Intelligence for Information Retrieval

Edited by
Dharmender Saini
Gopal Chaudhary
Vedika Gupta

CRC Press
Taylor & Francis Group
Boca Raton London New York

CRC Press is an imprint of the
Taylor & Francis Group, an **informa** business

First edition published 2022
by CRC Press
6000 Broken Sound Parkway NW, Suite 300, Boca Raton, FL 33487-2742

and by CRC Press
2 Park Square, Milton Park, Abingdon, Oxon, OX14 4RN

CRC Press is an imprint of Taylor & Francis Group, LLC

Names: Saini, Dharmender, editor. | Chaudhary, Gopal, editor. | Gupta, Vedika, editor.
Title: Computational intelligence for information retrieval / edited by Dharmender Saini,
    Gopal Chaudhary, Vedika Gupta.
Description: First edition. | Boca Raton : CRC Press, 2022. | Series: Sensors communication for
    urban intelligence | Includes bibliographical references and index.
Identifiers: LCCN 2021028865 (print) | LCCN 2021028866 (ebook) | ISBN 9780367680800 (hbk)
    | ISBN 9780367680831 (pbk) | ISBN 9781003134138 (ebk)
Subjects: LCSH: Expert systems (Computer science) | Artificial intelligence--Industrial
    applications. | Computational intelligence. | Information storage and retrieval systems.
Classification: LCC QA76.76.E95 C654 2022 (print) | LCC QA76.76.E95 (ebook) |
    DDC 006.3--dc23
LC record available at https://lccn.loc.gov/2021028865
LC ebook record available at https://lccn.loc.gov/2021028866

ISBN: 978-0-367-68080-0 (hbk)
ISBN: 978-0-367-68083-1 (pbk)
ISBN: 978-1-003-13413-8 (ebk)

DOI: 10.1201/9781003134138

Typeset in Palatino
by SPi Technologies India Pvt Ltd (Straive)

Contents

Preface vii
Editors ix
Contributors xi

1. Hybrid Computational Intelligence for Pattern Recognition 1
   Abhishek Bhatt, Vandana Thakur

2. Secure Image Transmission Using Nested Images 15
   Parul Arora, Shriya Khare, Nehal Joshi, and Bhaskar Chopra

3. Accist: Automatic Traffic Accident Detection and Notification
   with Smartphones 35
   *Vanita Jain, Eeshita Gupta, Manu S. Pillai, Priyanshi Bhola,
   and Gopal Chaudhary*

4. Emotion Prediction through EEG Recordings Using
   Computational Intelligence 47
   Asif Hasan, Azizuddin Khan, Asma Parveen, and Rajit Nair

5. Finger Vein Feature Extraction Using Contrast Enhancement
   Dynamic Histogram Equalization for Image Enhancement 63
   R. Anandha Jothi, G. Thenmozhi, J. Nityapriya, and V. Palanisamy

6. Song Recommendation Using Computational Techniques Based
   on Mood Detection 75
   *Madhav Kindra, Harshit Garg, Srishti Jhunthra, Vikrant Dixit, and
   Vedika Gupta*

7. Deep Learning Classification of Retinal Images for the Early
   Detection of Diabetic Retinopathy Disease 93
   N. Indumathi, B. Kalanjiyam, and R. Ramalakshmi

8. Protecting and Analyzing Big Data on Cloud Platforms 115
   Avita Katal, Niharika Singh, Vitesh Sethi, and Susheela Dahiya

9. Using Flutter to Develop a Hybrid Application of Augmented
    Reality 141
    Tarushee Kumar, Sudhanshu Sharma, Anirudh Sharma,
    Jatin Malhotra, and Vedika Gupta

10. Computational Intelligence Techniques for Recommendation
    System in Big Data 157
    Avita Katal, Vinayak Bajoria, and Vitesh Sethi

11. Predicting Melanoma Tumor Size through Machine Learning
    Approaches 197
    Dhruv Chadha, Nikita Jain, and Vedika Gupta

12. A Fuzzy-Based Approach for Characterization and
    Identification of Sentiments 219
    Madhav Kindra, Vikrant Dixit, and Vedika Gupta

13. Fingerprint Alterations Type Detection and Gender
    Recognition Using Convolutional Neural Networks and
    Transfer Learning 237
    Gaurav Kataria, Akansh Gupta, V. Sirish Kaushik,
    Gopal Chaudhary, and Vedika Gupta

14. Content-Based Image Retrieval Using Intelligent Techniques 257
    Prashant Srivastava, Manish Khare, and Ashish Khare

Index 275

Preface

This book will provide the readers with a comprehensive and recent exposition in the area of Computational Intelligence for Information Retrieval. This book will concentrate on computational intelligence approaches imitate human ways of processing information and mechanisms of reasoning as well as other biologically inspired processes such as evolution and collective intelligence found in many naturally occurring systems such as content-based image retrieval using intelligent techniques, hybrid computational intelligence for pattern recognition and intelligent innovative systems.

Also, this book will shed light on various applications of Big Data-facilitating industries to unearth hidden information in areas ranging from customer behavior to how their businesses function, providing vital and core insights that can affect the sustainability and profitability of an organization such as Protecting and Analyzing Big Data on Cloud Platforms. In a world where technology is advancing at a rapid pace, data-driven scientific discovery represents one of the most exciting developments – already making a huge impact in the social and services sectors enabled by the Internet of Things (IoT) and cloud computing.

We will include a wide range of issues – from the tools and languages of artificial intelligence to its philosophical implications and thus provides a plethora of theoretical as well as experimental research, along with surveys and impact studies. The information retrieval methods for text retrieval, multimedia information retrieval, data mining methods, text classification, and document clustering problems provide some really interesting findings. This book entails computational intelligence frameworks, approaches, models, and systems for efficient retrieval of information such as Using Flutter to develop a Hybrid Application of Augmented Reality, Computational Intelligence Techniques for Recommendation System in Big Data and A Fuzzy-based Approach for Characterization and Identification of Sentiments.

A wider selection of topics for this book is leading toward a thorough understanding of the reader for information retrieval basics and integration of computational intelligence with information retrieval. This book aims to investigate how computational intelligence frameworks are going to improve information retrieval systems. The emerging and promising state-of-the-art of human-computer interaction is the motivation behind this book. Further, the book aims to showcase the basics of information retrieval and computational intelligence for beginners as well as their integration and challenge discussions for existing practitioners.

Editors

Dharmender Saini is currently working as Principal at Bharati Vidyapeeth's College of Engineering, New Delhi, India. Dr. Saini is an active academician with an established background of working in the government administration industry, skilled in Patent Research. He holds PhD, M. Tech, and B. Tech in the field of computer science. He is a keen researcher and established professor in the field of cryptography. Dr. Saini has over 23 years of experience in research and academia. He served in industry for about 7 years actively involving himself in patent research, portfolio management, landscaping, and infringement analysis. He has been serving in academia for the last 23 years. He also has the experience of working as a consultant for Microsoft for 4 years and also served as consultant to Defense Research and development (DRDO). Dr. Saini is currently executing a project with Department of Science (DST), Indian Government) and research on drone-based city surveillance as principal investigator. He holds four Microsoft certifications including Infringement Analysis, US Patent Law, and claim charts and specification map. Dr. Saini is qualified as an Indian Patent Agent certified by Indian Government. He has organized many conferences and Hackathons to encourage entrepreneurship, prosper research, protect intellectual property (IP) and academics.

Gopal Chaudhary is currently working as an assistant professor in Bharati Vidyapeeth's College of Engineering, Guru Gobind Singh Indraprastha University, Delhi, India. He holds a Ph.D. in Biometrics from the Division of Instrumentation and Control Engineering, Netaji Subhas Institute of Technology, University of Delhi, India. He received a B.E. degree in Electronics and Communication Engineering in 2009 and a M.Tech. degree in Microwave and Optical Communication from Delhi Technological University (formerly known as Delhi College of Engineering), New Delhi, India, in 2012. He has 50+ publications in referred National/International Journals & Conferences (Elsevier, Springer, Inderscience) in the area of Biometrics and its applications. His current research interests include soft computing, intelligent systems, information fusion, and pattern recognition. He has organized many conferences and special issues.

Vedika Gupta is currently working as an assistant professor at Bharati Vidyapeeth's College of Engineering, New Delhi, India. Dr. Gupta is an active researcher with a demonstrated history of working in the government administration industry, skilled in Research and Development (R&D), data analysis, engineering, and programming. She holds PhD from the National

Institute of Technology Delhi (NITD) in the area of social media sentiment analysis. Dr. Gupta has 8 years of experience in research and academia. She has authored and co-authored more than 35 publications in various international SCI journals and conferences of high repute. Dr. Vedika is an active reviewer in several journals of Springer, IEEE, and Elsevier. She is a lifetime CSI member. Dr. Gupta heads Data Science Research Group, Bharati Vidyapeeth's College of Engineering, New Delhi, India. She is a member of Machine Intelligence Research Lab, Text Analytics Research Group at Banaras Hindu University, Varanasi, India.

Contributors

R. Anandha Jothi
Department of Computer Applications
Alagappa University
Karaikudi, Tamil Nadu, India

Parul Arora
Jaypee Institute of Information
    Technology
Noida, India

Vinayak Bajoria
Software Engineer
InsideView Technologies, Inc.
Hyderabad, India

Abhishek Bhatt
College of Engineering Pune
Pune, India

Priyanshi Bhola
Amity University
Noida, India

Dhruv Chadha
Bharati Vidyapeeth's College of
    Engineering
New Delhi, India

Gopal Chaudhary
Bharati Vidyapeeth's College of
    Engineering
New Delhi, India

Bhaskar Chopra
Jaypee Institute of Information
    Technology
Noida, India

Susheela Dahiya
Department of Computer Applications
School of Computer Science
University of Petroleum and Energy
    Studies
Dehradun, India

Vikrant Dixit
Bharati Vidyapeeth's College of
    Engineering
New Delhi, India

Harshit Garg
Bharati Vidyapeeth's College of
    Engineering
New Delhi, India

Akansh Gupta
Bharati Vidyapeeth's College of
    Engineering
New Delhi, India

Eeshita Gupta
Bharati Vidyapeeth's College of
    Engineering
New Delhi, India

Vedika Gupta
Bharati Vidyapeeth's College of
    Engineering
New Delhi, India

Asif Hasan
Psychophysiology Laboratory,
    Department of Psychology
Aligarh Muslim University
Aligarh, India

N. Indumathi
Department of Computer
  Applications
Kalasalingam Academy of Research
  and Education
Krishnankoil, Tamil Nadu, India

Nikita Jain
Bharati Vidyapeeth's College of
  Engineering
New Delhi, India

Vanita Jain
Bharati Vidyapeeth's College of
  Engineering
New Delhi, India

Srishti Jhunthra
Bharati Vidyapeeth's College of
  Engineering
New Delhi, India

Nehal Joshi
Jaypee Institute of Information
  Technology
Noida, India

B. Kalanjiyam
Department of Computer Science
  and Engineering
Kalasalingam Academy of Research
  and Education
Krishnankoil, Tamil Nadu, India

Avita Katal
Department of Virtualization,
  School of Computer Science
University of Petroleum and Energy
  Studies
Dehradun, India

Gaurav Kataria
Bharati Vidyapeeth's College of
  Engineering
New Delhi, India

Azizuddin Khan
Psychophysiology Laboratory,
  Department of Humanities and
  Social Sciences
Indian Institute of Technology
Bombay, India

Ashish Khare
Department of Electronics and
  Communication
University of Allahabad
Prayagraj, Uttar Pradesh,
  India

Shriya Khare
Jaypee Institute of Information
  Technology
Noida, India

Manish Khare
Dhirubhai Ambani Institute of
  Information and Communication
  Technology
Gandhinagar, Gujarat, India

Madhav Kindra
Bharati Vidyapeeth's College of
  Engineering
New Delhi, India

Tarushee Kumar
Bharati Vidyapeeth's College of
  Engineering
New Delhi, India

Jatin Malhotra
Bharati Vidyapeeth's College of
  Engineering
New Delhi, India

Rajit Nair
Jagran Lakecity University
Bhopal, India

J. Nityapriya
Department of Computer
  Applications
Alagappa University
Karaikudi, Tamil Nadu, India

V. Palanisamy
Department of Computer
  Applications
Alagappa University
Karaikudi, Tamil Nadu, India

Asma Parveen
Department of Psychology
Aligarh Muslim University
Aligarh, India

Manu S. Pillai
Bharati Vidyapeeth's College of
  Engineering
New Delhi, India

R. Ramalakshmi
Department of Computer Science
  and Engineering
Kalasalingam Academy of Research
  and Education
Krishnankoil, Tamil Nadu,
  India

Vitesh Sethi
Department of Virtualization
University of Petroleum and Energy
  Studies
Dehradun, India

Anirudh Sharma
Bharati Vidyapeeth's College of
  Engineering
New Delhi, India

Sudhanshu Sharma
Bharati Vidyapeeth's College of
  Engineering
New Delhi, India

Niharika Singh
Department of Informatics, School
  of Computer Science
University of Petroleum and Energy
  Studies
Dehradun, India

V. Sirish Kaushik
Bharati Vidyapeeth's College of
  Engineering
New Delhi, India

Prashant Srivastava
NIIT University, Neemrana
Rajasthan, India

Vandana Thakur
Technocrats Institute of Technology
Bhopal, India

G. Thenmozhi
Department of Computer
  Applications
Alagappa University
Karaikudi, Tamil Nadu, India

1

Hybrid Computational Intelligence for Pattern Recognition

Abhishek Bhatt

College of Engineering Pune, Pune, India

Vandana Thakur

Technocrats Institute of Technology, Bhopal, India

CONTENTS

1.1 Introduction 1
    1.1.1 Computational Intelligence and Hybrid Intelligence...................2
1.2 Evolution of Computational Intelligence in Health Care 3
    1.2.1 Artificial Neural Network 4
    1.2.2 Machine Learning 4
    1.2.3 Deep Learning 5
1.3 Hybrid Computational Intelligence for Disease Prediction 5
    1.3.1 Prediction of COVID-19 with Hybrid Computational
         Intelligence...6
    1.3.2 Analysis of Parkinson's Disease Using EEG Images 7
1.4 Areas of Hybrid Computational Intelligence for Future Research.......10
    1.4.1 Expert Systems 10
    1.4.2 Neural Nets...10
    1.4.3 Searching Process..11
    1.4.4 E-Learning...11
    1.4.5 Solving the Constraint..11
1.5 Conclusion 12
References...12

1.1 Introduction

The emergence of the Information Age has made a profound impact on health sciences. Data from various stages of health care organizations migrate across the different stages of these organizations. Intelligent machines help health practitioners in both the medical and administrative environments [1]. Studies have shown that these methods are rising in popularity because they can manage vast clinical data volumes and ambiguous details. Computational intelligence is based on biologically inspired algorithm computations [2]. Based on this area, there are main pillars such as neural networks [3], genetic algorithms [4], and fuzzy systems [5]. Neural networks can be used for function approximation problems and can classify artifacts. These artificial intelligence (AI) algorithms include supervised, unsupervised, and reinforcement learning. Genetic algorithms are search methods focused on genetic variants in natural systems. They depend on random and non-random genetic mutations. Populations are built over many generations in this case. It takes an evolutionary approach to solve broad complex problems [6]. Fuzzy logic is based on fuzzy set theory to allow reasoning which is fluid or approximate, as opposed to defined and precise. Fuzzy logical variables can accommodate "partial truth" in addition to the "true" and "false" values. Several types of computations have been used for solving world problems. Furthermore, fuzzy and genetic networks may also be used in the area of medical science [7].

1.1.1 Computational Intelligence and Hybrid Intelligence

Some technologies will have their benefits, while others will be sources of grief. In order to gain maximum success, creative and intelligent approaches are also essential. This mutualization should ensure there are no potential problems to worry about. It combines two smart strategies. By combining the neural networks with fuzzy techniques, the final solution is a mixture of both neural and neuro-fuzzy knowledge. When it comes to predicting future behaviors, fuzzy and neural networks are the building blocks of soft computing. AI is a type of computation focused on rationality and thinking. The system is composed of a neural network with a degree of human expert input to produce a neuro-creative system. They train a nation's citizens to be educated and employees to be prepared. A number of neurons are interconnected by way of running through the vectors. In more simple terms, the architecture consists of two neurons and the third layer consists of two neurons. Similarly, implicit rules were found by artificial neural networks (ANNs). This clarifies the actions of the neural expert system's thinking process when running on new data. This user-friendly interface is used to connect the neural expert system to the user, being built to allow their cooperation. It finally infers the data processing in the design and passes the knowledge to the neural information structure.

Expert systems rely upon logical reasoning combined with decision trees [10]. Neural networks use parallel processing and human brain structure to extract meaningful patterns.

- The future AI system will undergo a significant impact on the development and control of brain function.

- Knowledge in the rule-based expert system is cached as an "if-then" condition, whereas synaptic connections between neurons denote knowledge in ANNs.

- Expert device awareness can be separated into different laws that the expert can interpret and carry out. There is no distinct segment of information from which the weight of each synapse is learned. Layers of expertise are integrated with the whole network here. If synaptic weights shift in the brain, it may have uncertain and unpredictable effects.

- Neural networks and expert systems are claimed to be more capable than the current system by combining the two into one. A system of connected neural networks is referred to as a neural expert system.

1.2 Evolution of Computational Intelligence in Health Care

Modern health care technology is spreading across the world, helping people get healthier and live longer [11]. At first, the advancements have been driven by mobile devices' advent and the rising need for clinical record keeping due to medical progress. Computers today have a little more autonomous and advanced programming and applications. The advances in technology, especially ML [12] and AI applications, accelerate the speed of health technology change [12].

AI is one of the most effective technologies used in health care. Because medical data are becoming increasingly accessible and due to advances in big data, diagnostic technologies have complemented the potential of AI's existing use in the health care sector. The close relationship between medical problems and future AI techniques will enable many relevant data to be presented to decision-making [13] in health care. The applications of machine learning (ML) and AI in health care have allowed the field to tackle one of its key issues, which is the discovery of new drugs. Any other technical approach still faces problems. Most of the difficulties encountered in health care have been tackled by using AI technology, including regulatory aspects, patient and provider understanding, and data sharing. AI never reaches out to any of the listed challenges as it has shifted away from performing any of these tasks. The aim of AI and ML in the health care industry is to reshape the industry and make it possible. AI needs access to enough data to be useful in medicine. AI has made advancements in classifying complex variables faster and more accurately. Technologies of AI, including AI wellness apps, may allow people to assess their symptoms and take care of themselves whenever possible.

AI can help people become more independent and feel more dignified and comfortable at home. As AI is bound by data availability and data quality, it has some limitations. Also, a lot of computing power is required in the study of large and complex datasets. The clinical profession still requires social expertise that cannot be done by AI. The ML method assists in investigating such unstructured data as DNA, electro-physics, and medical images. ML uses knowledge-based computational methods to conclude. The data of the patients and their healing processes will be taken into consideration in ML algorithms.

A patient's "essence" essentially includes all information crucial to diagnosing their ailment. Simplifying complex data is ML support for AI. Significant groundbreaking advancement has occurred in this sub-field of neural networks. This has raised the overall interest in different fields of health and medication across substance and study area. Deep neural networks may diagnose complex health problems. When using AI techniques in health care-based applications, you can use any new modern mobile phone. AI can be used to combat essential health problems efficiently. AI will make a wrong judgment, and that will bring another critical issue. Who should be accountable when AI gets it wrong?

1.2.1 Artificial Neural Network

ANNs are among the most common ML models known in recent history. In a neural network, several computation nodes are organized in layers. Data flows through the network, being processed along the way. The network is trained to produce useful and predictable guesses by recognizing trends in a collection of labeled training data. During the neural network training, the network's parameters – the strength of neuron – are modified to recognize trends resulting in the training data. If the pattern is mastered, it can be used to guess data that are not seen before, such as generalizing to new data. It has long been recognized that ANNs are very versatile, but also that they are computationally difficult to program. This has, in turn, led researchers to concentrate on other ML models. ANNs are one of the leading techniques in ML nowadays. Since the growth of big data, several strong parallel processing devices (e.g., GPUs) have appeared, making the algorithms for neural networks faster. There is tremendous growth in ANN research as well, which pushes other functions of ML to progress.

1.2.2 Machine Learning

One develops and studies methods that help the machine learn from experiences. The aim is to construct mathematical models that can be given good input data and generate useful outputs. ML models are obtained by analyzing sets of training data and modified to generate accurate predictions by an optimization algorithm. The objective of the models is to generalize and produce accurate forecasts for unseen data. A model's generalization potential is usually estimated in a separate dataset, the validation dataset, and used to construct the model. After several iterations of training, the final model is checked on a test set to assess its accuracy. ML is loosely classified according to how a model processes the input data during preparation. A typical approach in reinforcement learning is to provide an agent that learns from their environment through trial and error while enhancing some objective function. AlphaGo and AlphaZero were recent implementations of reinforcement learning in ML [14]. In unsupervised learning, the machine is involved without any control to find patterns in the data. Many modern ML systems fit into the supervised learning group. In this way, the computer system can be used to perform tasks that need human intelligence. From a series of input-output examples, this ML model can accomplish specific data-processing functions. Here, image annotation using hand-labeled data is applicable because the image or medical image-related problem is addressed [15]. ML has a long history and is divided into various sub-fields and has deep understanding, getting the bulk of attention. Many surveys and overviews of deep learning are available online [8,9,16,18,19,26].

1.2.3 Deep Learning

Previously, ML models were designed for simple tasks and built on manually designed features. Using deep learning, computers learn useful representations directly from raw data. One of the more common forms of deep learning is the one that is developed using neural networks. The common feature of deep learning methods is their emphasis on learning the features of the data. The main difference between deep learning approaches is that it produces results faster. Work experience is incorporated into one problem; thus, it improves through the same training process. Study in medical imaging is driving the interest in convolutional neural networks (CNNs) [17]. Without a CNN, real-time language recognition usually had to be done by hand or performed by less efficient ML methods such as classifiers. When it became possible to use features learned directly from the data, many of the hand-trained image features became very useless [15]. The unique characteristic of most CNNs is that they are mostly context free. Let us now look at the building blocks of CNNs, TV News.

1.3 Hybrid Computational Intelligence for Disease Prediction

Health diagnosis refers to the method of assessing a person's condition based on symptoms and signs. The patient's specific details needed for disease diagnosis are gathered. This is a daunting treatment since it is

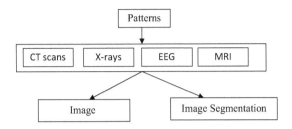

FIGURE 1.1
Computation intelligence approach.

difficult to diagnose. Thus, there needs to be a broad range of diseases. This incorporates an overview of statistical statistics. You can pick up useful knowledge, identification of patterns, and distinction among patterns. Figure 1.1 has shown the different types of patterns available during image segmentation.

1.3.1 Prediction of COVID-19 with Hybrid Computational Intelligence

This is according to the new studies in radiology imaging technology that has helped researchers uncover the virus CO19. AI and radiophotophysics will help make this diagnosis and at the same time relieve doctors' skills shortage. Automatic detection of chest X-ray patterns of CO19 is demonstrated. In the model DarkCAT, it has been implemented to provide correct binary classification as well as multi-class detection. The system applied the average classification's ability to distinguish between the two- and multi-class responses, respectively, to achieve a precision of 98.46 and an accuracy of 91. This complex network has added a total of 17 convolutionary layers and different filters to each layer. Using this technique, radiologists can confirm the initial reading on the spot and keep records in the cloud.

DarkCovidNet is used to develop a real-time object detection system. A means of real-time object detection has been built into this method. The model has designed with fewer layers and filters compared with the original Darknet. However, the typical DarkCovidNet architecture includes five max pool layers and 19 convolutional layers. Figure 1.2 depicts C with the Max pooling (also called convolution) layer and the Max pool. Complex neural networks with multiple layers have been used for object detection.

A subtler approach has been developed because they ran into a snag with classification. The model includes 17 convolutional layers, followed by one deep neural network layer then batch normalization, followed by Leaky ReLU activation. In the three consecutive layers, the forward and inverse lookups, all of the parameters have the same setup. To standardize the various inputs and improve stability, and speed up training, the dataset is

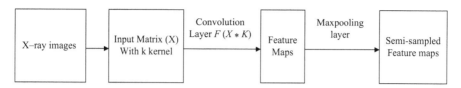

FIGURE 1.2
Computation intelligence approach.

TABLE 1.1

Computational Values of DarkCovidNet Model

Overlapped	Precision (%)	Sensitivity (%)	Specificity (%)	F1 Score (%)	Accuracy (%)
No detection	98.91	99.72	91.68	99.31	98.97
COVID-19	98.01	91.65	99.62	94.723	98.97
Average	98.46	95.685	95.65	97.0165	98.97

normalized. Deadly neurons use a slightly different version of the ReLU algorithm. The negative part of Leaky ReLU derivatives does not have large values, and Max pooling, a feature extraction network, is used. The filter reduces the size of the input by concealing the maximum percentage. This classification model has proven successful in the binary and multi-class setting. In type-O classification, it detects pneumonia, while in the multi-class, it labels the X-ray images. The model has used 1,748,352 parameters, and for weight updation, Adam optimizer has been used.

The five-fold validation method is used to calculate the classification model's efficiency for both binary and triple problems. Eighty-percent of the images are of the training set, and 20% of the images are used for validation. The splits have been folded to be used in the test execution stage. The model DarkCovidNet uses 100 epochs for training.

Above Table 1.1 has shown that the deep learning model called DarkCovidNet has shown promising results in binary and multi-class classification with the average accuracy of 98.97%.

1.3.2 Analysis of Parkinson's Disease Using EEG Images

Four neural network designs were developed to tackle the task of EO decoding: CNN, recurring neural network (RNN), 3D-CNN-RNN, and 2D-CNN-RNN [20]. The first model was the CNN [6] which served as a baseline model [21]. All four models used the raw information as inputs. This section describes how the training feedback and the model training strategy for these models should be represented. The three layers are joined into a CNN.

First of all, it is the convolution layer that interprets the input signal. Results are considered as a next level function. A pooling layer is usually used to pick features between two cooling layers, which decreases the entire map size. Both neurons are completely connected with neurons in a fully connected layer inside the pooling layer, and high-level functions are used to detect the input signal in various classes. RNN has a special computational property and can remember past knowledge without being reliant on the main memory. The model is based on a recurrent unit within GRU [22]. The steps used for constructing the architecture are as follows:

1. CNN with pooling-convolutional structure is used for auditory neuron mission. The kernel is the first layer that over time converges, and the kernel in the second layer performs electronic spatial filtering.

2. The model is able to validate that pitch perturbation contributes to greater neural responses in Parkinson's disease (PD) patients relative to healthy controls. To classify data into short-term memory storage, RNN was constructed through the division of raw data into 700 steps, with 64 components at each level (i.e., step length). The outputs of the last 20 phases were finally transferred through the fully connected layer and the softmax classification layer.

To explore the temporal significance of time-series data, the CNN structure is used to extract spatial information and the RNN structure is used due to their strong modeling capability. CNN is able to filter spatially, since it has one layer. It has 350 units, of which 40 are each (i.e., step length). The outputs of the last 50 steps of the structure of the RNN were fed into completely connected strata and subsequently into a classification layer.

A more advanced hybrid recurrent neural network has been developed to process spatial information. The CNN2D structure was built with two convolutionary layers in comparison with CNN. The kernel performs a time convolution in the first layer, the convolution process in the second layer is carried out in the 3D function map to extract spatial characteristics which implies that a second layer of convolution operates on pixel instead of color value. 3D-RNN is identical in the whole architecture to 2D-CNN-RNN (Figure 1.3).

The EEG can be depicted as topographically ordered pictures (i.e., voltage distributions across the scalp surface). Various studies have taken EEG images as a classifier input to CNN for processing those basic signals [23]. The reason for integrating spatial filters into EEG signals is to examine the energy patterns displayed at different locations of the brain. However, the hierarchical compositionality of local and global EEG modulations could scarcely be seen. Since EEG relies on the temporal hierarchy of local and world characteristics, CNN should be constructed with first-level local

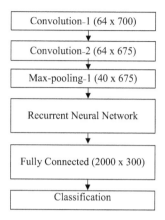

FIGURE 1.3
Architecture of D-CNN-RNN.

hierarchical characteristics and deeper-level globally hierarchical features. The input was then defined as a two-dimensional matrix with the number of occurrences (700) as the width and electrode number (64) as the height [24].

A similar result was obtained when regressing the EEG data into time to minimize the dimensionality of EEG data.

This study aimed to detect PD using EEG signals by deep learning ML models using two conventional deep learning architectures (CNN and RNN) and two hybrid convolutional-recurrent neural networks (3DCNN-RNN and 2D-RNN). The models significantly outperformed traditional models of AI. The research demonstrated the advantages of unsupervised learning algorithms for classifiers in dealing with raw EEG data in an automated manner, unique fashion, a point of view that may aid in clinical interpretation. Table 1.2 has represented the accuracy comparison of the different types of deep learning models discussed in this chapter.

TABLE 1.2

Comparison of Existing Deep Learning Model with the Hybrid Model

Model	Average Accuracy (%)
CNN	80.89
RNN	76.00
2D-CNN-RNN	81.13
3D-CNN-RNN	82.89

1.4 Areas of Hybrid Computational Intelligence for Future Research

In this section, we will discuss some of the other areas in which hybrid computational intelligence can be applied.

1.4.1 Expert Systems

Expert systems are the most used AI methods. Expert systems are in various software areas to assist in answering questions posted to them by users and finding solutions for those questions. It can be used to endorse choices in the field of medical science, in the investment world of finance, or online systems that vary from the most superficial technical diagnosis to complex, advanced hybrid approaches. An expert system encompasses a database of expertise relevant to the field of study. The "emptiness" of the database and the expert method, collectively as the "the shell," if something is full of information, must be loaded with energy; therefore, it should be full of knowledge. The shell should be able to sustain a knowledge base and interactive functions and integration into other applications. It means that you must first build an expert system by choosing or adapting an expert system and then adding expertise and information to your shell. The second stage is significantly more complicated and therefore requires a lot of time. One example of a system to use in cyberspace protection is a security strategy. A significant effort has been made in this area by the expert system to pinpoint security measures and use existing tools. We have also looked at approaches that use expert methods in intrusion detection that date back to the 1970s [25].

1.4.2 Neural Nets

Neural networks have a storied past, which began in the year 1957 with Frank Rosenblatt's discovery of "stereognosis" supervisory learning of binary classifiers (to paraphrase) (functions that decide whether the information conveyed by vector numbers belongs to a specific class). Artificial neurons are one of the most commonly used elements in neural networks together with a limited number of perceptron to learn and solve issues. Artificial neurons can be used in neural networks. ANNs could consist of a network of artificial neurons that can support a massive number of AIs. In the scheme of these networks, speed is the most influential characteristic feature. They work very well for pattern recognition, grouping, and in aphasia as well as for responding to responses. Applications in photography and graphic design may be made of both the digital and analog technologies.

Neural networks are also ideal for intrusion detection and prevention [27]. Scientific studies show how to use these networks for DoS detection [28], malware classification [29], computer worm detection [28], bot detection [30], spam detection [31], and automated forensic analysis [32] efficiently.

1.4.3 Searching Process

Most creative programs have search features, but these can be used in different ways; thus, useful search algorithms are critical to the program's overall success. Although meeting the criteria, additional knowledge will help in developing this research approach. AI has been applied in many ways as a search method. Dynamic programming has been used to tackle vulnerability and security issues but is not typically thought of as an AI task.

1.4.4 E-Learning

Learning expands or reorganizes the knowledge base by reorganizing the information structure or building an inference engine. All three aspects of this term are covered in the term ML heading. Problems in that are they do not fall into two broad categories of "solved" and "unsolved" (the complex values of symbolic education, such as learning the importance of some parameters and teaching concepts, language structures, functions, and even behavior learning). Both supervised and unsupervised learning use AI techniques. Supervised learning is especially beneficial when small quantities of data are scarce, but gathering a massive amount of log data in this fashion is typical of computer security. The art of unsupervised data discovery was pioneered in AI [33]. Parallel learning algorithms that are designed to be carried out on similar hardware have produced an excellent curriculum. This technique uses genetic algorithms and neural networks to look for creative solutions.

1.4.5 Solving the Constraint

Constraint solving involves using logical operators, tables, equations, or inequalities to assist in problem-solving when given a set of constraints. The answer to a question is a set of values that satisfy all conditions (constraints have a wide-ranging character; e.g., constraints on finite groups, functional constraints, rational trees). Most problems boil down to constraints on satisfaction at a deficient level. This is also due to the large number of calls that are needed in constraint programming, in decision support, as well as for computational reasoning.

1.5 Conclusion

Hybrid computer intelligence for pattern recognition is proposed in this chapter. Two different approaches to the diagnosis of medical imaging have shown hybrid computer intelligence. Hybrid information is a combination of at least two techniques. The first approach is based on hybrid CNN named DarkCovidNet which is used to classify the diseases based on COVID-19 and non-COVID-19. The next segment addressed the task of PD detection based on the EEG signal, with four architectures of deep learning algorithms – two conventions for deep learning (CNN and RNN) models and two hybrid convolutionary RNNs (2D-CNN-RNN and 3D-CNNRN). The results showed that hybrid models exceed conventional deep learning models. This chapter emphasizes the potential of deep learning algorithms which provide a new approach in promoting clinical decision making, for the automated classification of crude EEG data without handcraft features. For more studies on the correct diagnosis of illness, clustering would be extended to broad datasets.

References

[1] M. Teimouri, F. Farzadfar, M. Soudi Alamdari, et al. "Detecting diseases in medical prescriptions using data mining tools and combining techniques," *Iran J. Pharm. Res.*, 15 2016, 113–123.

[2] R. Iqbal, F. Doctor, B. More, S. Mahmud, and U. Yousuf, "Big data analytics: Computational intelligence techniques and application areas," *Technol. Forecast. Soc. Change*, 153, 2020, 119253. doi:10.1016/j.techfore.2018.03.024

[3] X. Jiang and R. E. Neapolitan, *Artificial Intelligence: With an Introduction to Machine Learning* (2nd ed.). Chapman and Hall/CRC, 2018.

[4] M. Kokkolaras, C. Audet and W. Hare, "Derivative-free and blackbox optimization. Springer series in operations research and financial engineering,".*Optim Eng.*, 20.3, 2019, 955–957. doi:10.1007/s11081-019-09422-9

[5] C. Grosan and A. Abraham, Fuzzy expert systems," in *Intelligent Systems*. Intelligent Systems Reference Library, vol. 17. Springer, Berlin, Heidelberg, 2011.

[6] R. Ligrone, *Biological Innovations that Built the World*, 2019, doi:10.1007/978-3-030-16057-9

[7] A. P. C. Chan, D. W. M. Chan, and J. F. Y. Yeung, Overview of the Application of 'Fuzzy Techniques' in Construction Management Research," *J. Constr. Eng. Manag.*, 2009, Published online 2009. doi:10.1061/(asce)co.1943-7862.0000099

[8] S. Ghosh, S. Biswas, D. Sarkar, and P. P. Sarkar, A novel Neuro-fuzzy classification technique for data mining,"*Egypt. Informatics J.*, 15.3, 2014, 129–147.

[9] S. Sahin, M. R. Tolun, and R. Hassanpour, "Hybrid expert systems: A survey of current approaches and applications," *Expert Syst. Appl.*, 39.4, 2012, 4609–4617.

[10] E. Bek-Pedersen, M. Lind, and B. A. Asheim, "AI based real-time decision making," in *Society of Petroleum Engineers – Abu Dhabi International Petroleum Exhibition and Conference 2019*, ADIP, 2019.

[11] G. Hinton, "Deep learning-a technology with the potential to transform health care," *JAMA*. 2018.

[12] F. Jiang et al., "Artificial intelligence in healthcare: Past, present and future," *Stroke Vascular Neurol.*, 2.4 2017.

[13] N. T. Hung and K. L. Yen, The role of motivation and career planning in students' decision-making process for studying abroad: A mixed-methods study," *Rev. Argentina Clin. Psicol.*, 29.4, 2020, 252–265.

[14] D. Silver et al., "Mastering the game of Go without human knowledge," *Nature*, 550.7676, 2017, 354–359.

[15] A. Esteva et al., "Dermatologist-level classification of skin cancer with deep neural networks," *Nature*, 542.7639, 2017, 115–118.

[16] Y. Lecun, Y. Bengio, and G. Hinton, "Deep learning," *Nature*, 521.7553, 2015, 436–444.

[17] S. Haykin and B. Kosko, "GradientBased learning applied to document recognition," in *Intelligent Signal Processing*, 2010.

[18] T. Ozturk, M. Talo, E. A. Yildirim, U. B. Baloglu, O. Yildirim, and U. Rajendra Acharya, "Automated detection of COVID-19 cases using deep neural networks with X-ray images," *Comput. Biol. Med.*, 121 2020, 103792.

[19] R. Nair, S. Vishwakarma, M. Soni, T. Patel, and S. Joshi, "Detection of COVID-19 cases through X-ray images using hybrid deep neural network," *World J. Eng.*, 2021.

[20] X. Shi, T. Wang, L. Wang, H. Liu, and N. Yan, "Hybrid convolutional recurrent neural networks outperform CNN and RNN in Task-state EEG detection for parkinson's disease," in *2019 Asia-Pacific Signal and Information Processing Association Annual Summit and Conference, APSIPA ASC 2019*, 2019.

[21] A. Khan, A. Sohail, U. Zahoora, and A. S. Qureshi, "A survey of the recent architectures of deep convolutional neural networks," *Artif. Intell. Rev.*, 2020.

[22] Z. Che, S. Purushotham, K. Cho, D. Sontag, and Y. Liu, "Recurrent Neural Networks for Multivariate Time Series with Missing Values," *Sci. Rep.*, 2018.

[23] P. Bashivan, I. Rish, M. Yeasin, and N. Codella, "Learning representations from EEG with deep recurrent-convolutional neural networks," in *4th International Conference on Learning Representations, ICLR 2016 – Conference Track Proceedings*, 2016.

[24] R. T. Schirrmeister et al., "Deep learning with convolutional neural networks for EEG decoding and visualization," *Hum. Brain Mapp.*, 2017.

[25] I. Kotenko and A. Ulanov, "Multi-agent framework for simulation of adaptive cooperative defense against internet attacks," in *International Workshop on Autonomous Intelligent Systems: Multi-Agents and Data Mining*. Springer, Berlin, Heidelberg, 2007.

[26] Y. Freund and R. E. Schapire, "Large margin classification using the perceptron algorithm," *Mach. Learn.*, 1999.

[27] N. El Kadhi, K. Hadjar, and N. El Zant, "A mobile agents and artificial neural networks for intrusion detection," *J. Softw.*, 7.1, 2012, 156–160.

[28] I. Ahmad, A. B. Abdullah, and A. S. Alghamdi, "Application of artificial neural network in detection of DOS attacks," in *SIN'09 – Proceedings of the 2nd International Conference on Security of Information and Networks* (pp. 229–234), 2009.

[29] M. Shankarapani, K. Kancherla, S. Ramammoorthy, R. Movva, and S. Mukkamala, "Kernel machines for malware classification and similarity analysis," in *Proceedings of the International Joint Conference on Neural Networks*, 2010.

[30] P. Salvador, A. Nogueira, U. França, and R. Valadas, "Framework for zombie detection using neural networks," in *Proceedings – 2009 4th International Conference on Internet Monitoring and Protection, ICIMP 2009*, 2009.

[31] C. H. Wu, "Behavior-based spam detection using a hybrid method of rule-based techniques and neural networks," *Expert Syst. Appl.*, 36.3, 2009, 4321–4330.

[32] B. K. L. Fei, J. H. P. Eloff, M. S. Olivier, and H. S. Venter, "The use of self-organising maps for anomalous behaviour detection in a digital investigation," *Forensic Sci. Int.*, 162, 2006, 33–37.

[33] D. Heckerman, "A tutorial on learning with Bayesian networks," *Innovat. Bayesian Netw.*, 2008, 33–82.

2

Secure Image Transmission Using Nested Images

Parul Arora, Shriya Khare, Nehal Joshi, and Bhaskar Chopra
Jaypee Institute of Information Technology, Noida, India

CONTENTS

2.1 Introduction 15
2.2 Literature Survey 17
2.3 Proposed Work 18
    2.3.1 Phase 1: Creation of Intermediate Transformed Image 18
    2.3.2 Phase 2: Creation of Final Transformed Image 19
    2.3.3 Phase 3: Recovery of Intermediate Transformed Image 19
    2.3.4 Phase 4: Recovery of Confidential Image 19
2.4 Idea for Enhancing the Transformed Image Security 21
    2.4.1 Intermediate Transformed Image Creation 21
    2.4.2 Final Transformed Image Creation 24
2.5 Recovery Process 27
2.6 Experimental Results and Discussion 27
    2.6.1 Technique 1 27
    2.6.2 Technique 2 28
    2.6.3 Comparison between Both Techniques 29
    2.6.4 Comparison of Transformed Image Creation Results
        with Other Techniques 31
    2.6.5 Conclusion 31
Conflict of Interest 32
References 32

2.1 Introduction

Currently, in a world ruled by digital media, people rely on sharing information through images, videos, texts, and many of these media are confidential. There have been a lot of methods for securing such digital media in the past using steganography and many other related tools [1, 2]. The military information exchange consists of sending and receiving images that can be accessed while they are being transmitted. Another field that has grown exponentially is health care. One of the major developments is the swift movement of test results from labs to the hospitals, ensuring immediate actions especially during emergencies. These test results, often images, are combined with the patient information which is in turn confidential. This information exchange can prove to be dangerous [3] if done in an unsafe manner, or through unreliable channels of transmission that may give access to illegal recipients through simple manipulations on them. Apart from the general steganography methods [4], there have been many other image encryption methods [5] that make use of the properties of images such as redundancy and correlation and hide the image by converting it into another image which has a lot of noise. This may attract an illegal operator's attention due to the humongous degree of randomness in the image. There are many techniques to achieve the goal of enshrouding the confidential image. Concealing the lowest bits of the confidential content into the bits of noisy images is one such method [6, 7] while converting or replacing the statistical profile of the file that needs to be concealed with that of another file, known as mimic functions being another. One of the most common methods of performing image-based steganography is hiding the bits of the confidential content into random data that increases the randomness and the redundancy of the files and many more. A method for increasing the safety of images during their transmission has been proposed, in which the confidential image that is to be transmitted is transformed into another unrelatable image, meaning a transformed image, that looks highly similar to a preselected cover image, combining several image processing techniques and steganography [8]. Motivated by this approach and to provide additional security, the authors have devised a new approach in this work. This approach is developed further to improve certain aspects, mainly the security, by introducing an additional cover image, that presents an additional layer for the transformed image to be hidebound. Thus, the objective of this paper is to introduce a technique that would be beneficial for hiding private or classified images and would not arouse any attacker's attention due to a negligible amount of noise and randomness in the transformed image.

The paper is organized in the following sections: Section 2.2 gives the literature survey of the work. Section 2.3 briefs the proposed work by dividing the approach into different phases. The methodology of coding and decoding the images is thoroughly explained in Section 2.4 and Section 2.5. Experimental results have been discussed in Section 2.6 and the last section gives the conclusion.

2.2 Literature Survey

Any image can be treated as a confidential image, the image which is to be safeguarded and another image can be taken as the cover image to hide it. The confidential image can be transformed into the cover image using the proposed techniques. Many researchers have worked in this field and have proposed the ideas, which includes reversible contrast mapping [9], and even by using a common framework that will generate the digital stegano medium [10]. It is primarily focused on protecting files having rich text format by modifying the inter-character spacing. Separate algorithms for compression and steganography allowing more data to be hidden can also be used [6], including the use of masking to embed useful information in digital media, and using Lempel-Ziv-Welch technique for compression of the media. However, the compression of the digital media often results in the loss of useful information in bits and pieces, resulting in the redundancy while recovering it. The most general methods of information hiding can be by increasing the redundancy [11] or by correlating neighbouring pixels [12], but these methods result in a considerable amount of distortion and hence are not feasible. A different approach was looked upon for the purpose of security enhancement [13], in which the horse-step algorithm has been used. It relies on the amount of redundancy with the cover media to hide the secret data. Modifying the redundancy to hide information depends highly on the power of human vision. Authors [14] have hidden the information in images by changing the appearance of the picture instead of changing its features. The information is obscured using the appearance of multiple other images, placed together. This approach [15] presents a very specific application, wherein any input image is transformed into an ancient image using different simulation engines in a dimensional environment. Multiple mosaic formation techniques have been surveyed in [16]. Adaptive pixel pair matching along with an asymmetric cryptographic algorithm was used [17], converting pairs of pixels into coordinates and then searching their neighbourhood to find an appropriate place to hide the message. However, the message is pre-processed using Rivest-Shamir-Adleman algorithm, and modifying the secret image is in turn very risky. Text steganography was focused upon [18], explaining several ways of applying steganography to hide confidential text messages using embedding and word mapping, but the proposed ideas cannot be applied to hiding images and other digital media thereby narrowing the spectrum of its use. A small text message was hidden in two bitmap images using basic-level steganography [19] but clearly points out the use of only selective images for the purpose of hiding the message. Steganography is compared with symmetric algorithms and cryptography [20]. The vulnerabilities of classical watermarking towards geometrical distortions are pointed out [21], and different watermarking schemes have been proposed as solutions. A brief yet useful comparative study between cryptography, steganography, and watermarking is carried out [22]. Video steganography has been carried out using the hamming code [23], and steganography using the hypertext markup language has also been explored as a possibility [24]. Yet another way of performing steganography has been proposed by detecting the features of images, particularly edges [25], but might prove to be difficult to carry out if the images vary a lot in terms of the features used. Image steganography and hiding capacities in different formats of images have been discussed.[8] Another approach for hiding a text message inside images has been performed [26], and the storage capacity of the cover media is increased using compression of the important text message to be hidden.

2.3 Proposed Work

This work accentuates a safe method to transmit and recover a confidential image through a channel. Thus, the proposed work can be divided into the following phases:

2.3.1 Phase 1: Creation of Intermediate Transformed Image

After the selection of confidential image, a cover image, according to the sender's choice, is selected. Then using a combination of image processing and steganography techniques, including the division of both the images into blocks, calculation of average standard deviation for each of the red, green, and blue pixels in all the blocks of both the images, mapping those blocks together, and then performing least significant bit substitution in the end for the transformation, the confidential image is reconstructed into the intermediate transformed image as shown in Figure 2.1.

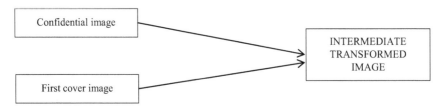

FIGURE 2.1
Creation of intermediate transformed image.

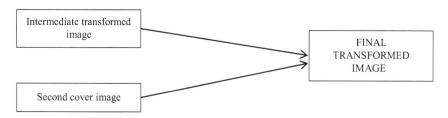

FIGURE 2.2
Creation of final transformed image.

2.3.2 Phase 2: Creation of Final Transformed Image

The intermediate transformed image is then reconstructed using another cover image, which is selected purely by the sender as shown in Figure 2.2.

2.3.3 Phase 3: Recovery of Intermediate Transformed Image

This is the first stage of recovery. The bits embedded within the final transformed image consist of the bits from the second cover image and the intermediate transformed image.

These bits are then extracted, and the intermediate transformed image is recovered as shown in Figure 2.3.

2.3.4 Phase 4: Recovery of Confidential Image

The bits within the intermediate transformed image consist of the bits from the confidential image and the first cover image. They are hence extracted to recover the confidential image, thereby completing the recovery process as shown in Figure 2.4.

In this way, this work will set forth a method to hide more than one image into another image making a transformed image that can go undetected.

The procedure explained above can be applied while using two images as cover images, the first and the second cover image. Using more than one cover images adds on to the already high level of security for hiding

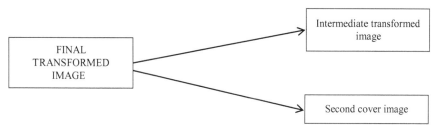

FIGURE 2.3
Recovery of intermediate transformed image.

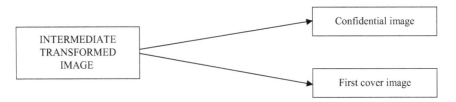

FIGURE 2.4
Recovery of confidential image.

the confidential image. The novel approach of using two cover images to hide a single confidential image has been explained in the following block diagram in Figure 2.5. The initial embedding stage consists of the use of steganography and least significant bit substitution resulting in the formation of the intermediate transformed image, which again undergoes embedding along with the second cover image to form the final transformed image.

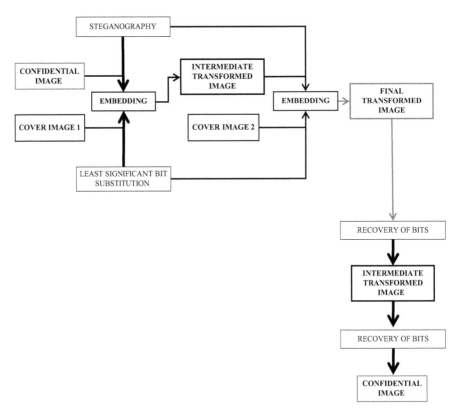

FIGURE 2.5
Block diagram of the proposed work.

The intermediate transformed image is recovered from it, and then the confidential image is recovered from the intermediate transformed image.

2.4 Idea for Enhancing the Transformed Image Security

This section explains the new course of action that has been undertaken. The level of security while using one cover image was already strong, but it can further be enhanced by adding an additional stage of embedding using a second cover image. Figure 2.6 shows the embedding phase, with the first stage being the creation of the intermediate transformed image and the second stage consisting of the creation of the final transformed image.

2.4.1 Intermediate Transformed Image Creation

The first stage consists of the transformation of the confidential image into an intermediate transformed image. This task is accomplished by performing a series of methods using the concepts of image processing and steganography. In the beginning of the procedure, both the images are divided into an equal number of blocks. Each pixel in each block is processed for the calculation of their attributes, mainly the standard deviation. The values of the standard deviation depend upon the amount of red, green, and blue colours present in each pixel. The standard deviation is calculated for the red, green, and blue values of all the pixels separately. There would be a lot of such values, since there are a lot of pixels in each block of an image. Hence, to represent each block of an image, arithmetic mean of all the standard deviation values is calculated. This value of the average standard deviation represents each block and is used in relating the confidential and the cover blocks. The values of all the representative numbers are then sorted and stored dynamically into separate arrays for both the confidential and the cover images. The sorted arrays are then compared together, and the closest values that can be related are then exchanged. This method of relating the blocks on the basis of the colour values of both the images is called mapping. A mapping sequence

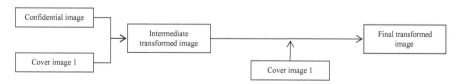

FIGURE 2.6
Creation of the final transformed image in two stages.

is obtained. This sequence represents the relation between both the images, showing the blocks in each of the images that resemble with each other. The pixel values are converted to binary values. After that, using the technique of least bit substitution, the last bits of each pixel of the cover image 1 are replaced by the first of the confidential image. By the end of the first stage, we obtain the intermediate transformed image.

Figures 2.7 and 2.8 depict the confidential and the cover image being divided into blocks, respectively. The pixels of the confidential blocks are represented by {P1, P2, … Pn} and {P'1, P'2, …, P'n}.

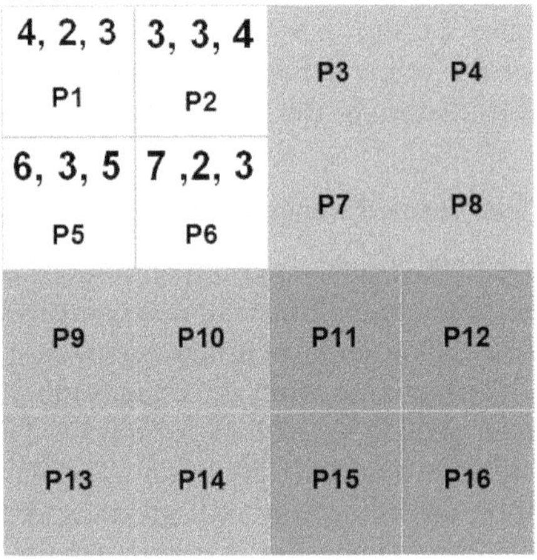

FIGURE 2.7
Confidential image divided into confidential blocks.

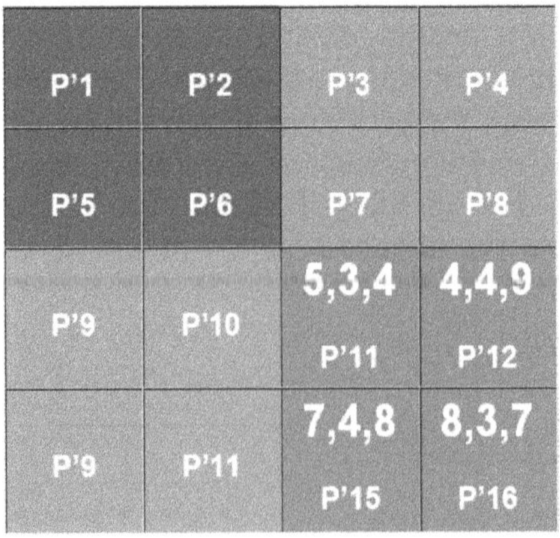

FIGURE 2.8
Cover image divided into cover blocks.

For each pixel, there are red green and blue values which are used for the calculation of the standard deviation and the arithmetic mean to find the average standard deviation that represents each block for both the confidential and the cover image.

The standard deviation for each colour in each pixel is calculated using Equation 2.1.

$$SD = \sqrt{\frac{\sum(x - \bar{x})^2}{n}} \tag{2.1}$$

The average standard deviation (mean of all the standard deviations in a block) is calculated using Equation 2.2.

$$avg(S.D) = \frac{S.D1 + S.D2 + \ldots + S.Dn}{n} \tag{2.2}$$

In Figure 2.9, the values of the average standard deviations of the confidential and cover images are mapped. The values 2.18 and 1.93 are the closest and hence mapped together.

In the same way, the landscape images are selected for the proposed work, and they are divided into confidential and cover blocks in Figures 2.10 and 2.11 shows the result obtained after mapping the confidential and cover image blocks, before converting the values into bits and performing LSB substitution.

The intermediate transformed image shown in Figure 2.13 is obtained using the confidential image Figure 2.12(a) and the first cover image Figure 2.12(b). The procedure of LSB substitution has been followed here, in which the least significant bits of the first cover image are replaced by the most significant bits of the confidential image. In the same way, when two cover images are

FIGURE 2.9
Mapping image blocks into cover blocks in accordance to the average deviation.

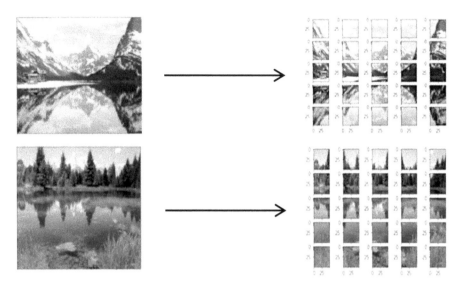

FIGURE 2.10
Depiction of selected landscape image into blocks.

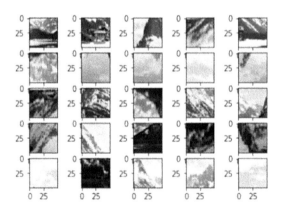

FIGURE 2.11
Result obtained after mapping.

taken, the process of the nested LSB substitution has been explained in the following Section 2.4.2 and depicted in Figure 2.14.

2.4.2 Final Transformed Image Creation

The intermediate transformed image is treated as a confidential image, and a second cover image, in the block diagram explained in Figure 2.5, is used

FIGURE 2.12
(a) Confidential image and (b) first cover image.

FIGURE 2.13
Intermediate transformed image.

for the transformation. The same procedure is then carried out, and the final image obtained is the final transformed image. The most significant bits of the confidential image and cover image are chosen from the intermediate transformed image and these replace the least significant bits of the second cover image.

The pixels of the final transformed image contain the information of the intermediate transformed image. Hence, it can be recovered from the latter. Similarly, the confidential image can be recovered from the intermediate transformed image.

As shown below, Figures 2.15 through 2.17 represent results after all the intermediate stages that are carried out in the creation of the final transformed image.

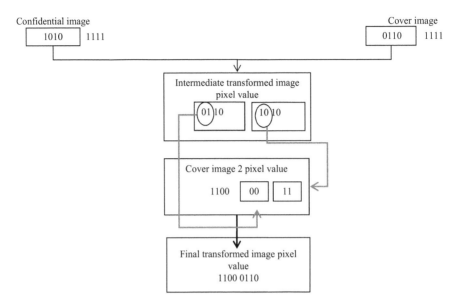

FIGURE 2.14
Sequence and order of least bit substitution during transformed image creation.

(a)   (b)

FIGURE 2.15
(a) Confidential image and (b) first cover image.

(a)   (b)

FIGURE 2.16
(a) Intermediate transformed image and (b) second cover image.

FIGURE 2.17
Final transformed image.

2.5 Recovery Process

The images are recovered by modifying and recovering the bits that were substituted before. For instance, each of the pixel bit values in the final transformed image consists of the bits from the intermediate transformed image and the second cover image. Hence, these bits are extracted and zeroes are padded to complete the eight-bit sequence in the first stage of recovery. Similarly, each of the pixel bit values in the intermediate transformed image consists of the bits from the first cover image and the confidential image, which are extracted and separated, resulting in the final recovery of the confidential image.

The zeroes padded do account for some loss, but that loss is negligible because the image recovered is highly similar to the confidential image. This is mainly because of the information embedding method that uses each pixel for storing the information of our confidential image, and the number of pixels being very high reduce the loss to a very meagre value. This procedure is represented in Figure 2.18.

2.6 Experimental Results and Discussion

2.6.1 Technique 1

The first technique proposed here consists of the confidential image being hidden within a cover image, without any constraints on the degree of similarity. The differences between the cover image and the transformed image, and between the recovered image and the confidential image, have been calculated using the root mean square error between them.

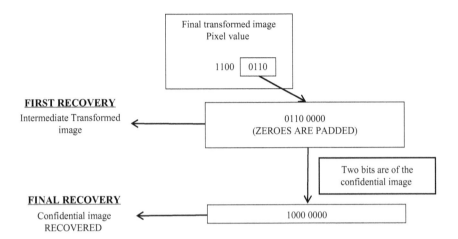

FIGURE 2.18
Recovery process.

FIGURE 2.19
Recovered image from the intermediate transformed image.

Thus, the root mean square error is used as an evaluation metric. The final recovered image obtained from technique 1 is shown in Figure 2.19.

The results obtained during the transformation and the recovery stages by changing the dimensions of the image using technique 1 have been shown in Table 2.1.

2.6.2 Technique 2

This technique involved the usage of two cover images to make the confidential image hidebound. The first stage consists of transforming the confidential image into one of the cover images, and then this intermediate transformed

TABLE 2.1

RMSE Results Obtained Using Technique 1

Variation in Image Dimensions	Stage	RMSE Value	
		Block Size: 50*50	Block Size: 25*25
250*250*3	Transformation	6.75699341	6.727795
	Recovery	8.8717944	8.8717944
500*500*3	Transformation	6.807871	6.84699
	Recovery	8.878371	8.87837
1000*1000*3	Transformation	6.87348	6.90570
	Recovery	8.867672	8.867672

(a) (b)

FIGURE 2.20

(a) Recovered intermediate transformed image and (b) recovered confidential image.

image is further reconstructed into the second cover image, thereby increasing the security immensely. The evaluation metric used was the root mean square error between the image after the final transformation and one of the cover images. Both the cover images were taken randomly again, stating no constraints over the degree of similarity between them and the confidential image. The final transformed image has been presented in Section 2.4.2. The recovered intermediate transformed image and the recovered confidential image are shown in Figures 2.20(a) and (b), respectively.

The results obtained during the transformation and the recovery stages by changing the dimensions of the image using technique 2 have been shown in Table 2.2.

2.6.3 Comparison between Both Techniques

From Table 2.3, it is evident that the proposed method proved out to be exceptional from both concerns: security and accuracy. Depending on the

TABLE 2.2

RMSE Results Obtained Using Technique 2

| Variation in Image Dimensions | Stage | RMSE Value | |
		Block Size: 50*50	Block Size: 25*25
250*250*3	Transformation	6.639310	6.630575
	Recovery	35.664556	35.664556
500*500*3	Transformation	6.624528	6.69277
	Recovery	35.664654	35.664654
1000*1000*3	Transformation	6.711441	6.7229281
	Recovery	35.622545	35.622545

application and use of the method, one of the two methods can be chosen accordingly.

Both the approaches represent a trade-off between accuracy and security. If the application requires commendable accuracy while transmitting the confidential image, technique 1 can be used, because it gives exceptionally low error values for both the transformation and the recovery stages. In an application which prioritizes on security, technique 2 would be more suitable. As stated earlier, there is a trade-off between the accuracy and security. Hence, in order to increase the security during the transformation stage, the recovery stage accuracy decreases meagrely.

However, since the priority of the application lies in increasing the security, the second technique gives a better efficiency during the transformation stage. The confidential image is transformed twice in the second technique, and the results obtained certainly prove that it is more efficient than technique 1 during the transformation stage.

TABLE 2.3

A Comparison between Both the Techniques

Variation in Image Dimensions	Stage	RMSE Value			
		Technique 1		Technique 2	
		Block Size		Block Size	
		50*50	25*25	50*50	25*25
250*250*3	Transformation	6.756993	6.727795	6.639310	6.630575
	Recovery	8.871794	8.871794	35.664556	35.664556
500*500*3	Transformation	6.807871	6.84699	6.624528	6.69277
	Recovery	8.878371	8.87837	35.664654	35.664654
1000*1000*3	Transformation	6.87348	6.90570	6.711441	6.7229281
	Recovery	8.867672	8.867672	35.622545	35.622545

TABLE 2.4

Comparison of Proposed Techniques with Other Techniques

Techniques	MOSAIC Image [27]	MOSAIC Image [28]	Proposed Method
RMSE Values	47.651	33.935	**6.727**

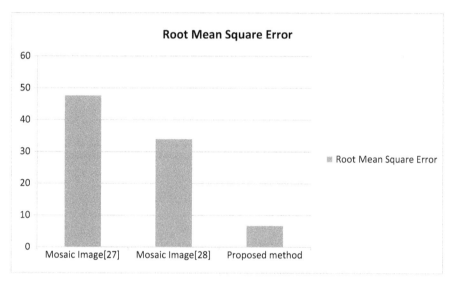

FIGURE 2.21
Comparison of root mean square error across different techniques.

2.6.4 Comparison of Transformed Image Creation Results with Other Techniques

The following Table 2.4 consists of the results while creating the mosaic image reported in [27] and [28] using a set of personal cover and confidential images, showing the ineffectiveness of the technique followed due to the high amount of root mean square error between the confidential image and the transformed image. Through the graph in Figure 2.21, it can be inferred that there is a considerable difference among the RMSE obtained using our approach and the ones obtained using other techniques.

2.6.5 Conclusion

An unprecedented method for secure image transmission has been proposed, which not only hides confidential images with least distortion but also focuses on a feasible recovery. The confidential image can be transformed into a cover image of the user's choice which does not consist of any restriction other than being from the same colour palette. The confidential and the cover images are mapped using the statistical information, which is followed by an effective use of least significant substitution, resulting in the formation of an intermediate image. The recovery stage is then carried out to derive the confidential image back from the cover image. In this work, two approaches have been proposed, out of which initial approach uses one cover image and the another one involves nesting of multiple cover images. Through extensive experimentation, it has been proved that nested approach is far better for security concerns. Future work might include applying the same objective to other forms of digital media, like using other audio files as cover and confidential media along with images, hence expanding the application of this method.

Conflict of Interest

The authors declare that they have no conflict of interest.

References

[1] A.P. Westfeld, "Attacks on Steganographic Systems", Pfitzmann A. (eds) *Information Hiding, IH 1999*, Lecture Notes in Computer Science, 1768, 61–76. Springer, Berlin, 1999 [doi:10.1007/10719724_5].

[2] K. Bailey, K. Curran, "An evaluation of image based steganography methods using visual inspection and automated detection techniques", *Multimed. Tools Appl.*, **31**, 327, 2006 [doi:10.1007/s11042-006-0047-x].

[3] H.J. Baur, U. Engelmann, F. Saurbier, A. Schröter, U. Baur, H.P. Meinzer, "How to deal with security issues in teleradiology", *Comput. Methods Programs Biomed.*, **53**(1), 1–8, May 1997 [doi:10.1016/S0169-2607(96)01798-1].

[4] S. Sharda, S. Budhiraja, "Image Steganography: A review", *Int. J. Emerging Technol. Adv.Eng.*, **3**, January 2013.

[5] M. Nosrati, et al. "An introduction to steganography methods – TI Journals", *World Appl.Program.*, **1**, 191–195, August 2011.

[6] Y. Yiğit, M. Karabatak, "A stenography application for hiding student information into an image", *7th International Symposium on Digital Forensics and Security (ISDFS)*, 1-4, Barcelos, Portugal, 2019 [doi:10.1109/ISDFS.2019.8757516].

[7] A. Tripathi, J. S. Arya Nair, "Steganography using segmenting mosaic images with embedding data by 3 LSB S-type scanning", *Int. J. Latest Trends Technol.* **2**(4), July 2013.

[8] N. Hamid, Y. Abid, R. B. Ahmad, O. Al-Qershi, "Image steganography techniques: An overview", *Int. J. Comput. Sci. Security*, **6**, 168–187, 2012.

[9] D. Coltuc, J.-M. Chassery, "Very fast watermarking by reversible contrast mapping," *IEEE Signal Process. Lett.*, **14**(4), 255–258, April 2007 [doi:10.1109/LSP.2006.884895].

[10] V. Ganeshkumar, R.L.W. Koggalage, "*Secured communication using Steganography Framework with sample RTF implementation*", *2009 International Conference on Industrial and Information Systems (ICIIS)*, Sri Lanka, 2009 [doi:10.1109/ICIINFS.2009.5429897].

[11] N. Hamid, A. Yahya, R. Badlishah Ahmad, O. Al-Qershi, Dheiaa Aldeen Najim Alzubaidy, L. Kanaan, "Enhancing the Robustness of Digital Image Steganography Using ECC and Redundancy", *J. Inform. Sci. Eng.*, 2013.

[12] G. Swain, S. K. Lenka, "Pixel value differencing steganography using correlation of target pixel with neighboring pixels", *2015 IEEE International Conference on Electrical, Computer and Communication Technologies (ICECCT)*, 1–6, Coimbatore, 2015 [doi: 10.1109/ICECCT.2015.7226029].

[13] R. Poudyal, S. Shakya, "Performance analysis of stegano data with improved LSB substitution using horse step algorithm and advanced encryption standard", *International Conference on Advances in Computing, Communication Control and Networking (ICACCCN)*, 307–311, Greater Noida (UP), India, 2018 [doi:10.1109/ICACCCN.2018.8748723].

[14] M. Shirali-Shahreza, S. Shirali-Shahreza, "Collage Steganography", *5th IEEE/ACIS International Conference on Computer and Information Science and 1st IEEE/ACIS International Workshop on Component-Based Software Engineering, Software Architecture and Reuse (ICIS-COMSAR'06)*, 316–321, Honolulu, 2006 [doi:10.1109/ICIS-COMSAR.2006.27].

[15] S. Battiato, G. Puglisi, "3D ancient mosaics", *Proc. ACM Int. Conf. Multimedia*, 1751–1753, Florence, Italy, October 2010 [doi:10.1145/1873951.1874289].

[16] Battiato, G. Di Blasi, G. M. Farinella, G. Gallo, "Digital mosaic framework: An overview", *Eurograph.-Comput. Graph. Forum*, **26**(4), 794–812, December 2007 [doi:10.1111/j.1467-8659.2007.01021.x].

[17] F. Nizar, F. Latheef, A. Jamal, "RSA based encrypted data embedding using APPM", *International Journal of Engineering Trends and Technology (IJETT)*, **9**, 777–782, March 2014 [doi:10.14445/22315381/IJETT-V9P347].

[18] A. Koluguri, S. Gouse, P. Bhaskara Reddy, "Text steganography methods and it's tools", *International Journal of Advanced Scientific and Technical Research*, **2**(4), March–April 2014.

[19] M. Atallah Al-Shatnawi, "A new method in image steganography with improved image quality", *Applied Mathematical Sciences*, **6**, 3907–3915, 2012.

[20] K. Patel, S. Utareja, H. Gupta, "Information hiding using least significant bit steganography and blowfish algorithm", *International Journal of Computer Applications*, **63**, 24–28, 2013 [doi:10.5120/10527-5510].

[21] P. Bas, J. Chassery, B. Macq, "Geometrically invariant watermarking using feature points", *IEEE Transactions on Image Processing*, **11**(9), 1014–1028, September 2002, [doi:10.1109/TIP.2002.801587].

[22] V. H. Desai, "Steganography, cryptography, watermarking: A comparative study", *Journal of Global Research in Computer Sciences*, **3**, 33–35, December 2012.

[23] R. J. Mstafa, K. M. Elleithy, "A highly secure video steganography using Hamming code (7, 4)", *IEEE Long Island Systems, Applications and Technology (LISAT)*, 1–6, Farmingdale, 2014 [doi: 10.1109/lisat.2014.6845191].

[24] S. Mahato, D. K. Yadav, D. A. Khan, "A Modified approach to text steganography using HyperText markup language", *2013 Third International Conference on Advanced Computing and Communication Technologies (ACCT)*, 40–44, Rohtak, India, 2013 [doi:10.1109/ACCT.2013.19].

[25] N. Jain, S. Mesh Ram, S. Dubey, "Image steganography using LSB and edge–detection technique", *Int. J. Soft Comput. Eng. (IJSCE)*, 2012.

[26] V. Sharma, S. Kumar, "A new approach to hide text in images using steganography", *Int. J. Adv. Res. Comput. Sci.*, **3**, 701–708, 2013.

[27] J. Lai, W. H. Tsai, "Secret-fragment-visible mosaic image—A new computer art and its application to information hiding", *IEEE Trans. Inform. Forensic Security*, **6**(3), 936–945, September 2011 [doi:10.1109/TIFS.2011.2135853].

[28] Y. Lee, W. Tsai, "A new secure image transmission technique via secret-fragment-visible mosaic images by nearly reversible color transformations", *IEEE Trans. Circuits Syst. Video Technol.*, **24**(4), 695–703, April 2014 [doi:10.1109/TCSVT.2013.2283431].

3

Accist: Automatic Traffic Accident Detection and Notification with Smartphones

Vanita Jain, Eeshita Gupta, and Manu S. Pillai
Bharati Vidyapeeth's College of Engineering, New Delhi, India

Priyanshi Bhola
Amity University, Noida, India

Gopal Chaudhary
Bharati Vidyapeeth's College of Engineering, New Delhi, India

CONTENTS

3.1 Introduction 35
3.2 Related Work and Our Contribution 37
    3.2.1 Related Work 37
    3.2.2 Our Contribution 38
3.3 Research Methodology 39
    3.3.1 Data Pre-Processing 39
    3.3.2 Feature Extraction 40
    3.3.3 Proposed Model Architecture for Critical Accident
          Detection 41
    3.3.4 Mobile Alert and Cloud Computing 42
3.4 Conclusion 44
3.5 Future Work 44
References 45

3.1 Introduction

Everyone is terrified as soon as they hear the word accident but it has impact only for some days, thus following the rules and then going back to unsafe and rogue driving. One of the reasons for irresponsible driving is lack of patience,

DOI: 10.1201/9781003134138-3

which is more common than common sense in individuals, and yet another reason is competitive driving, i.e., driving not to reach the destination but to defeat the other driver and win against them. Competitive driving most of the time leads to a fatal accident claiming the life of the driver or those who are on the road. Winning against someone while indulged in competitive driving is considered as an act of pride and honor by youngsters. With this mentality, there are very limited options to make driving safe for drivers and the ones on the road.

Accidents are becoming the root cause for the devastating experience that parents have to go through upon losing their child. Everyone has felt this way at some point of their life and losing a loved one can completely devastate an individual. At various instances where accidents become an inevitable situation, every precaution and safety measure become unambiguously futile, gradually resulting in fatal circumstances. Accidents are something which cannot always be prevented or avoided but the mortality rate due to an accident can be reduced up to an appreciable rate by reducing the duration of time between the event of an accident and the arrival of the first help. By reducing the reaction time of help, we could effectively move a step toward saving a life.

Technical advancements can be observed in mobile industries at a significant rate as compared to the automobile industry. Today we have mobiles with various security features, e.g., panic mode, emergency mode, etc. [24–26]. While driving a vehicle has still not improved much, just the comfort of driving has increased but not the safety of the driver and passengers. Damage that can be caused by a failure in a mobile phone is remarkably lower than due to a car accident, yet the focus of the automobile industry giants is toward providing comfort for the compromise in the safety of the occupants. Innovating a specific technology for this purpose is impractical as it will remarkably increase the manufacturing cost and therefore limiting the cars available with this technology. Thus, using the technology that is already present in most cars can be a cost-effective method and will assist to make this technology available to maximum people. At the initial stage, even if we could save a single life, it would be a great achievement for us as we not only saved that one life but also the happiness of all the people connected with that person. Though it would seem an insignificantly small step for safety but at least it would trigger a series of activities for the safety of people and minimizing the road accidents.

This paper is composed of the following sections: the next section contains some related work in this field and our contribution to this field. In Section III, Research Methodology is discussed followed by a conclusion.

3.2 Related Work and Our Contribution

3.2.1 Related Work

The knowledge of the situation and how structures can be built appropriately to serve the situation were a subject of complex analysis. Different situational consciousness studies in the area of urgent medical dispatch (EMD) have been published a lot in recent years. Research on one of the world's biggest emergency systems [15] was performed by Blandford and William Wong. Situational awareness was established and exploited, especially among the senior EMD operators known as allocators. Even before anyone else notifies them by the smart AI systems which process CCTV images, these allocators can be notified [16].

Recent research of detecting vehicle accidents has been extremely conspicuous and has brought papers with excellent correctness as delineated in the outcomes published [21].

Chenyi Chen et al. propose to map an input image to a small number of key perception indicators that directly relate to the affordance of a road/traffic state for driving. Their representation provides a set of compact yet complete descriptions of the scene to enable a simple controller to drive autonomously [14].

Shunsuke Kamijo et al. developed an algorithm Markov Random Field (MRF) for intersection traffic images with the help of which they were able to tackle the tracing problem. Their algorithm was able to track and segment occluded with an accuracy of 93–96%, which led to the development of Hidden Markov Model (HMM) which not only recognizes accidents but also jamming and passing. Though the cases of accidents were less, still their algorithm was not able to predict with good accuracy [1].

Sang Jun, Wu Zhongyuan, Guo Pei, Hu Haibo, Xiang Hong, Zhang Qian, and Cai Bin [17] showed variations of improved YOLOv2 for object detection to improve feature extraction using multi-layer feature fusion strategy, and the repeated convolution layers in high layers were removed. Similar work was seen by Wangpeng He, Zhe Huang, Zhifei Wei, Cheng Li, and Baolong Guo [18] which used TF-YOLO for real-time object detection.

A System for the Autonomous Removal of False Detection Boxes [19] was done by Zhiyuan Lin, Qingxiao Wu, Shuangfei Fu, Sikui Wang, Zhongyu Zhang, and Yanzi Kong. Some of these techniques can be used to remove false alarms.

Md. Syedul Amin et al. utilized speed of the vehicle to detect accidents. They utilized a GPS receiver to monitor speed and compare with previous speed every second using a microcontroller unit. It assumes that an accident has occurred when the speed of the vehicle is below a certain speed. It sends the location captured using GPS if an accident occurs [2]. YongKul Ki et al. suggested a system for automatically detecting, recording, and reporting accidents at intersections by which they were able to determine the cause of accidents and intersection's features that impact safety. They achieved a really low false alarm rate of 0.34×10^{-6}% [3]. Jorge Zaldivar et al. proposed an android application which uses OnBoard Diagnostics (OBD-II) to monitor vehicles. They estimated GForce experienced by passengers in case of frontal collision. Details of the accident are sent via email or SMS to some already defined destinations. The app was able to react in less than 3 seconds when tested on an actual vehicle [4].

Jules White et al. proposed an accident detection model that combined sensor data such as accelerometer and acoustic information and context data which included determining whether the user is inside the car or not. They actually developed an accident detection prototype WreckWatch, an open-source android application [5].

3.2.2 Our Contribution

Considering the discussed points and related work in the previous section, the paper focuses on exhibiting an easy approach with the existing technology to provide quicker assistance and prevent fatality rate due to accidents.

We have used Transfer learning. Transfer learning is a quantitative modeling methodology for a specific but in several ways similar problem, which can then be used partially or entirely to boost training and enhance the efficiency of a model with respect to the topic of interest.

It involves, in deep learning [20], using weights in one or more layers from a pre-trained model of a network in a new model and keeping weights set, changing them thoroughly or modifying the weights fully during model testing.

This approach is implemented on convolutional neural networks (CNNs) [22] with ResNet pre-trained models [27–29]. The discussed method is in line with machine learning classification techniques of object detection. It uses feature extraction and gradually evolves itself from raw labeled data. The dataset is modified with noise to match the real-life scenarios of actual CCTV footage which may be blurred and have a lower frame quality [23].

We feed pairs of inputs and corresponding labeled scenarios of the intensity of the accident to the neural network. The aim is to predict any accident and the extent and severity of the same. We also employ threshold value for probability below which is not considered an accident; this helps to reduce false alarms.

Deploying the trained model to cloud [30] can be easily used further to embed the same to CCTV footage. This can trigger on time alerts to concerned authorities or provide immediate help.

3.3 Research Methodology

3.3.1 Data Pre-Processing

Accident-Images-Analysis-Dataset is being used for the study. Dataset includes a total of 10480 images for comprehensive analysis. Of total, 4898 images are classified as damaged and undamaged vehicles. In addition, the dataset classifies another 2636 vehicle images on the basis of their weight, i.e., heavy-duty, light-duty, or motorcycle (2-wheel). Another 2936 affected vehicles are further classified on the basis of accident severity, i.e., "low dangerous," "moderately dangerous," and "extremely dangerous." Figure 3.1

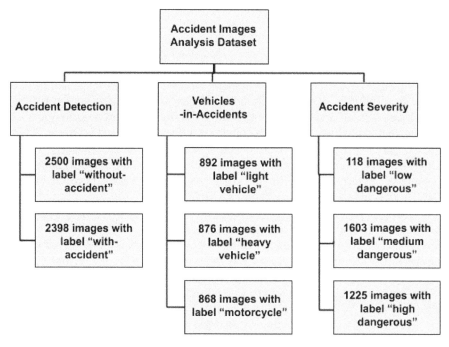

FIGURE 3.1
Description of accident-images-analysis-dataset.

provides the full overview of the dataset with the number of images in different categories.

The data are preprocessed before giving it as the input to the model. First, we have converted all the images into grayscale to reduce computational complexity, then resized the images to size 224 × 224, which is the input size of the feature extractor we have used. Most of the images in the database used were directly scraped from the web. As a result, a larger portion consisted of high-quality images. To increase generalization of the model, after scaling down those images into desired size, we have added nominal Gaussian noise to the scaled-down images to mimic inputs taken directly from low-resolution CCTVs. Such a procedure has shown a significant increase in accuracy at inference time.

$$b_x = \sigma\left(t_x\right) + c_x \tag{3.1}$$

$$b_y = \sigma\left(t_y\right) + c_y \tag{3.2}$$

$$b_w = p_w e^{t_w} \tag{3.3}$$

$$b_h = p_h e^{t_h} \tag{3.4}$$

3.3.2 Feature Extraction

Deep learning is a branch of artificial intelligence-based learning algorithms, which use several layers similar to that of the brain system to slowly generate high-level outputs, which are considered deep organized learning. There are a number of architectures including the recurrent neural network (RNN), DBN, CNN, which are used in numerous fields such as natural language processing or system traduction, analysis of medical images, computer vision, etc. There is also a range of different architectures.

A CNN as shown in Figure 3.2 uses a 3D tensor like an image, assigns learned weights and distinctions to various features, and trains them to differentiate one from the other. A ConvNet's architecture is close to that of the neurons in the human brain and has been influenced by the Visual Cortex organization.

The first step in our methodology is extracting the bounding boxes of cars. We have used YOLOv2 pretrained weights for the same [6]. You look only once (YOLO) is a real-time object detection system [13]. Unlike other object detection systems that remodel classifiers or localizers to perform detection, predictive high-scoring regions of the image are used to detect.

However, YOLO divided a single image into regions and bounding boxes to define probabilities for each region using a single deep neural learning

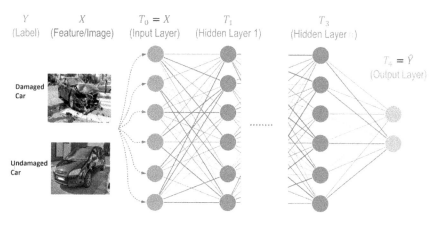

FIGURE 3.2
A CNN network.

network for each image. These bounding boxes are weighted by the expected probabilities. This particular approach makes it perfect for this kind of research.

Although for our usage, the original 80 class model is not necessary but for ease of implementation, we have used the default model in our work. If required, a separate model can be trained using these weights for localizing cars. Once the bounding boxes are acquired, we extract car images from the input stream and produce it as input to our CNN classifier. The CNN classifier takes an image of a car and classifies whether the car is damaged (critically) or not.

For this CNN architecture, we have used a pretrained ResNet-50-v2 model as a feature extractor [7, 12]. It is a CNN of 50 layers deep with a network of over one million images from the ImageNet database [11]. The network has learned the rich features that make it reliable. This network can classify images into a number of categories of objects, such as cars, trees, street lights, and many animals. We also used our dataset to classify new images using this ResNet-50 model.

3.3.3 Proposed Model Architecture for Critical Accident Detection

Our approach to critical accident detection consists of two components: the car image acquisition and classifying them into critically damaged or not. We have used YOLOv2 architecture for car image acquisition and trained our own CNN classifier for the latter.

The CNN classifier is trained using ResNet-50-v2 as the feature extractor [8]. Each of these has over 1,00,000 neurons. Then, we have removed the

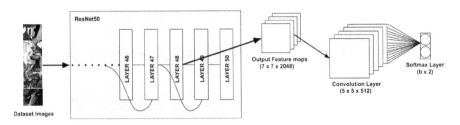

FIGURE 3.3
Architecture of the trained classifier.

last two convolutional layers of the model and added a convolutional layer
of kernel size 5 × 5 followed by a dense layer of two SoftMax activation
units (Figure 3.3 shows the architecture of the classifier). We use SoftMax
activation function, also known as soft argmax or normalized exponential function, to normalize our output. It maps the non-normalized output
layer and normalizes it to a probability distribution consisting of K probabilities proportional to the exponentials of the input numbers. The scope
is from (0,1).

The SoftMax function is defined by the following formula:

$$\sigma(z)_i = \frac{e^{z_i}}{\sum_{j=1}^{K} e^{z_j}} \text{ for } i = 1,\ldots,K \tag{3.5}$$

$$z = (z_1,\ldots z_K) \in R^K \tag{3.6}$$

The model is trained at a learning rate of 0.01 with exponential decay and
cross entropy loss for 40 epochs producing an F1-score of 0.78 at inference. To
suppress any true negatives or false alarms, we have introduced a threshold
and the system only triggers an alert if the probability of the vehicle being
crashed is above this threshold [9].

3.3.4 Mobile Alert and Cloud Computing

Our proposed system consists of one more component, i.e., a mobile application, which alerts the user/concerned authority about any accidents that
occurred in a connected CCTV.

AI as a cloud platform service makes intelligent automation easier for
users who are not interested in process complexity. The application does not
require technical knowledge. This would boost even the opportunity to help
the ones in need with cloud servers to a significant degree. High-capacity

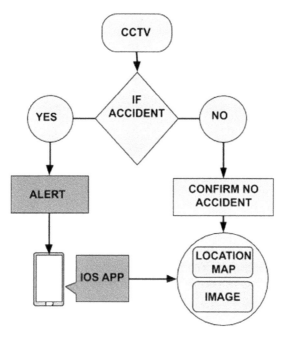

FIGURE 3.4
Block diagram of the working system.

networks, low-cost computers and storage devices, widespread use of virtual hardware, service-based architecture, and autonomous computing minimize response times and help to rapidly deliver real-time performance at low hardware costs.

A CCTV pushes the video footage to a cloud server in real time where our trained model is evaluating the incoming streams. Once the model detects a crashed car with a probability > 0.7, the system adds an entry to Firebase [10]. The mobile application picks up this entry and alerts the user (Figure 3.4 shows the block diagram of the system).

The entry consists of time and location of the CCTV along with the image that triggered the alarm. Such a provision decreases the chances of false alarms as the user can reassure the prediction before taking further steps.

Blue color symbolizes everything being normal and red color is by default the symbol of the danger in Figure 3.5. Similarly, when there is no abrupt change in the course of vehicle, the picture with blue color will be delivered by the system and as soon as the impact of any kind is detected an alert will be sent to an iOS app, depicted by the red background image, with information including the GPS location of the vehicle and an image of the damaged vehicle to have some human intervention.

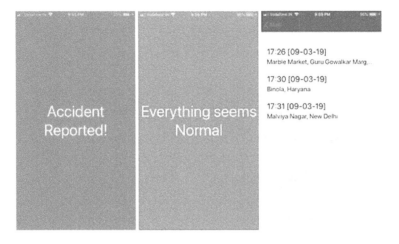

FIGURE 3.5
Screenshots of the mobile application.

3.4 Conclusion

We have introduced a combination of YOLO and CNNs for real-time accident detection systems with an F1 score of 0.78. YOLO is state-of-the-art and faster than other detection systems across a variety of detection datasets. Furthermore, it can be run at a variety of image sizes to provide a smooth tradeoff between speed and accuracy. Exploiting the performance of such a model, we have used a CNN for classification of the segmented car images.

3.5 Future Work

In future work, we hope to use similar techniques along with incorporating tracking capabilities in the model. We also plan to improve our detection results using more powerful matching strategies for assigning weak labels to classification data during training. Computer vision is blessed with an enormous amount of labeled data. We will continue looking for ways to bring different sources and structures of data together to make stronger models of the visual world.

References

[1] Kamijo, S., Matsushita, Y., Ikeuchi, K., & Sakauchi, M. (2000). Traffic monitoring and accident detection at intersections. *IEEE transactions on Intelligent Transportation Systems* 1(2):108–118.

[2] Amin, M.S., Jalil, J., & Reaz, M.B.I. (2012, May). *Accident detection and reporting system using GPS, GPRS and GSM technology*. In *2012 International Conference on Informatics, Electronics & Vision (ICIEV)* (pp. 640–643). IEEE, Dhaka, Bangladesh.

[3] Ki, Y.K., Kim, J.W., & Baik, D.K. (2006, August). *A traffic accident detection model using metadata registry*. In *Fourth International Conference on Software Engineering Research, Management and Applications (SERA'06)* (pp. 255–259). IEEE, Seattle, WA.

[4] Zaldivar, J., Calafate, C.T., Cano, J.C., & Manzoni, P. (2011, October). *Providing accident detection in vehicular networks through OBD-II devices and Android-based smartphones*. In *2011 IEEE 36th Conference on Local Computer Networks* (pp. 813–819). IEEE, Bonn, Germany.

[5] White, J., Thompson, C., Turner, H., Dougherty, B., & Schmidt, D.C. (2011). Wreckwatch: Automatic traffic accident detection and notification with smartphones. *Mobile Networks and Applications* 16(3):285–303.

[6] Viola, P., & Jones, M. (2001). Robust real-time object detection. *International Journal of Computer Vision* 4(34–47):4.

[7] Papageorgiou, C.P., Oren, M., & Poggio, T. (1998, January). *A general framework for object detection*. In *Sixth International Conference on Computer Vision* (IEEE Cat. No. 98CH36271) (pp. 555–562). IEEE, Bombay, India.

[8] Szegedy, C., Ioffe, S., Vanhoucke, V., & Alemi, A.A. (2017, February). *Inception-v4, inception-resnet and the impact of residual connections on learning*. In *Thirty-First AAAI Conference on Artificial Intelligence*, San Francisco, CA.

[9] Wu, Z., Shen, C., & Van Den Hengel, A. (2019). Wider or deeper: Revisiting the resnet model for visual recognition. *Pattern Recognition* 90:119–133.

[10] Moroney, L. (2017). Firebase cloud messaging. In *The Definitive Guide to Firebase* (pp. 163–188). Apress, Berkeley, CA.

[11] ImageNet n.d. http://www.image-net.org

[12] He, K., Zhang, X., Ren, S., & Sun, J. (2016). *Deep residual learning for image recognition*. In *Proceedings of the IEEE Conference on Computer Vision and Pattern Recognition* (pp. 770–778). Las Vegas, NV.

[13] Redmon, J., Divyala, S., Girshick, R., & Farhadi, A. (2016). *You only look once: Unified, real-time object detection*. In *Proceedings of the IEEE conference on computer vision and pattern recognition* (pp. 779–788).

[14] Chen, C., Seff, A., Kornhauser, A., & Xiao, J. (2015). *Deepdriving: Learning affordance for direct perception in autonomous driving*. In *Proceedings of the IEEE International Conference on Computer Vision* (pp. 2722–2730).

[15] Blandford, A., & William Wong, B.L. (2004). Situation awareness in emergency medical dispatch. *International Journal of Human-Computer Studies* 61(4):421–452.

[16] Champion, H.R., Augenstein, J., Blatt, A.J., Cushing, B., Digges, K., Siegel, J.H., & Flanigan, M.C. (2004) Automatic crash notification and the URGENCY algorithm: Its history, value, and use. *AEN* 26(2):143.

[17] Sang, J., Wu, Z., Guo, P., Hu, H., Xiang, H., Zhang, Q., & Cai, B. (2018). An improved YOLOv2 for vehicle detection. *Sensors* 18(12):4272.

[18] He, W., Huang, Z., Wei, Z., Li, C., & Guo, B. (2019). TF-YOLO: An improved incremental network for real-time object detection. *Applied Science* 9(16):3225.

[19] Lin, Z., Wu, Q., Fu, S., Wang, S., Zhang, Z., & Kong, Y. (2019). Dual-NMS: A method for autonomously removing false detection boxes from aerial image object detection results. *Sensors* 19(21):4691.

[20] Goodfellow, I., Bengio, Y., & Courville, A. (2016). Goodfellow-et-al-2016, Deep Learning, MIT Press. http://www.deeplearningbook.org

[21] Kumeda, B., Fengli, Z., Oluwasanmi, A., Owusu, F., Assefa, M., & Amenu, T. (2019). *Vehicle Accident and Traffic Classification Using Deep Convolutional Neural Networks*. In *2019 16th International Computer Conference on Wavelet Active Media Technology and Information Processing*, Chengdu, China (pp. 323–328). doi:10.1109/ICCWAMTIP47768.2019.9067530.

[22] Koushik, J. (2016). Understanding convolutional neural networks.

[23] Lin, Y. (2018). *Evaluation methods and applications in image recognition based on convolutional neural networks. Journal of Physics: Conference Series*, IOP Publishing 1087(6), p. 062023. Pingtung, Taiwan.

[24] Yasaswini, L., Mahesh, G., Siva Shankar, R., & Srinivas, L.V. (July 2018). Identifying road accidents severity using convolutional neural networks. *International Journal of Computer Sciences and Engineering* 6(7):354.

[25] Zhu, L., Guo, F., Krishnan, R., & Polak, J.W. (2018). *A deep learning approach for traffic incident detection in urban networks.* In *21st IEEE International Conference on Intelligent Transportation Systems (ITSC) Maui*, November 4–7. Maui, Hawaii.

[26] Zheng, M., Li, T., & Zhu, R. (March 2019). Traffic accident's severity prediction: a deep learning approach-based CNN network. *IEEE Access* 7:39897–39910.

[27] Krizhevsky, A., Sutskever, I., & Hinton, G.E. (2012). ImageNet classification with deep convolutional neural networks. *Advances in Neural Information Processing Systems*, 25, 1097–1105.

[28] McDonnell, M.D., & Vladusich, T. (2015). *Enhanced image classification with a fast-learning shallow convolutional neural network.* In *2015 International Joint Conference on Neural Networks (IJCNN)*, 2015.

[29] Zhao, J. (2016). Research of substation monitoring image recognition approach based on convolutional neural networks, Doctoral dissertation.

[30] Velte, T., Velte, A., & Elsenpeter, R.(2009). *Cloud Computing, A Practical Approach* (1st ed.). McGraw-Hill, New York.

4

Emotion Prediction through EEG Recordings Using Computational Intelligence

Asif Hasan
Aligarh Muslim University, Aligarh, India

Azizuddin Khan
Indian Institute of Technology, Bombay, India

Asma Parveen
Aligarh Muslim University, Aligarh, India

Rajit Nair
Jagran Lakecity University, Bhopal, India

CONTENTS

4.1 Introduction 47
    4.1.1 Electroencephalography 49
4.2 Literature Survey 50
4.3 Emotion Prediction of EEG Recordings through Computational
    Intelligence Methods 51
    4.3.1 Data Collection 51
    4.3.2 Data Preprocessing 52
    4.3.3 Computational Intelligence with Feature Selection
        and Feature Extraction 53
    4.3.4 Classification 55
    4.3.5 Evaluation Parameters 55
4.4 Conclusion 58
References 59

4.1 Introduction

There is interest in the automated recognition of human emotions in multimedia systems and brain-computer interfaces. For example, video games know the emotional state of the user. The computer is programmed to respond to the data. Emotions over such forms of content may be used to serve content to interesting users [1]. Emotions may be calculated using two-dimensional or three-dimensional dimensions of valence, arousal, and dominance. Many emotion models are two dimensional, and the subjective valence of this message is positive-negative. The second dimension is a measure of pleasure level. By looking at the public's speech and gestures, one can appreciate their emotions. The chapter will focus on emotion detection using physiological scales, which can produce more reliable measures of emotional states. And we focus on analyzing emotions, particularly emotional sounds and visuals. The topics covered in the chapter are as follows:

1. Computational intelligence on predicting the emotions [2];
2. Feature extraction and classification [3];
3. Construction of brain activity using computational intelligence methods;
4. Classification and performance evaluation.

The capacity to perceive human emotions plays a critical role in interpersonal relationships. Automatic emotion detection has been a significant subject of research since the early days. Also, because of this, there are numerous innovations in this area [4]. To express emotions, one uses voice, body language, and facial expressions. Emotion extraction and interpretation are essential to human-machine interaction and the feelings enrich human life. Non-verbal cues such as hand and body movements, the sound of the voice used to communicate emotion and provide input, and facial expression are all components of human interpersonal communication. Expressing feelings occur as people communicate with each other regularly. Interaction can be made richer if one understands and knows how to respond to the people's language.

Several studies on various approaches to understanding emotions from expression were performed [5]. Many tools were placed in place to assign the different states to vocalizations. To be understood, feelings need to be recognized by facial expressions. While the earliest scientific study on facial expressions dates back to the time of Aristotle, research on facial expressions can be traced to as early as the Aristotelian period. There has been active research into facial expression recognition since the early 1990s [6]. Actions and gestures on the face can be parameterized using muscle movements. Once these parameters have been defined, they can be used to communicate different facial expressions.

It is not easy to understand someone's emotions because they are based on meaning. One individual might pretend to be happy by smiling, but this is not because they are genuinely excited on the inside. Previously, machine learning approaches used to analyze emotion have earned attention because of facial expression, gesture, and voice recognition. Still, understanding someone's emotional state involves additional knowledge that is not found in facial expressions or movements and voice tone. As a result, electroencephalography (EEG) tests have gained popularity because the EEG test generates brain activity, giving rise to a signal that researchers, physicians, and other professionals may use to detect, diagnose, and help prevent both brains and mental diseases [7]. Emotion recognition is also supported by access to massive EEG datasets and advances in BCIs, providing more excellent reliability and accuracy.

EEG research has been applied widely in neuroscience and other related areas [8]. Machine learning algorithms have been used to discover the necessary knowledge for brain characteristic recognition and cognitive neuroscience studies [9]. The abundance of broad datasets and innovations in machine learning has significantly contributed to the implementation of deep learning architecture in analyzing EEG signals and the knowledge it can provide as a measure of brain functionality. An essential aspect of brain-computer interfaces is the development of the robust automated classification of these signals as shown in Figure 4.1.

4.1.1 Electroencephalography

To track the electrical activity of the brain, the use of EEG is needed. Invasive electrodes are also used, as in electrocorticography [11], which is often called intracranial EEG.

Ionic current inside the brain's neurons produces voltage fluctuations, and an EEG calculates these fluctuations. Although EEG recording refers to taking measurements of brain activity over time, it is only done by electrodes mounted on the scalp. Event-related potentials and spectral quality of EEG are the two major themes in diagnostic applications. Stimulus onset refers to when the action starts, such as when a button is pressed. The second study of brain waves, which are classified as brain waves since they oscillate, investigates various forms of oscillations (which are usually called "brain waves") that can be found in EEG signals in the frequency domain [12].

EEG is used in studies involving neurology, neuroscience, and biomedical engineering (e.g., brain-computer interfaces, BCI, sleep monitoring, and

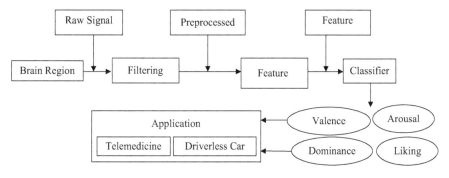

FIGURE 4.1
Computational intelligence approach for the EEG classification [10].

seizure detection) [13]. With these completely automatic, anyone could use EEG at home without background knowledge from a specific sector. In the traditional EEG classification pipeline, there are removing objects, extraction of features, and classification. The EEG database comprises a 2D (time and channel) matrix of valid values that reflect brain-generated potentials. The highly organized design of EEG data makes it ideal for machine learning. Multiple machine learning and pattern recognition algorithms have been implemented on the EEG data. Commonly used approaches are independent component analysis (ICA) [14], principal component analysis (PCA) [15], local Fisher's discriminant analysis (LFDA) [16], linear discriminant analysis (LDA) [17], support vector machines (SVM) [18], decision trees [19], canonical correlation analysis (CCA) [20], and others (SSVEPs).

Neural networks have been reasonably under-utilized compared to other machine learning approaches because of related practical problems, such as long computing time, vanishing or exploding gradients, etc. [21]. With a wide range of data available and recent developments of graphic processing units, neural network researchers explored different deep learning architectures (neural network architectures containing at least two hidden layers). These innovations have led to an exponential increase in the development of learning algorithms in the past decade. Yes, DNN improves performance in a wide range of traditionally tricky topics. Neural networks are generally believed to be less in need of special knowledge about the data. This gives significant advantages in the field of medical image analysis. Considerable progress has been made in EEG signals, particularly in the realm of machine learning.

4.2 Literature Survey

Advanced emotion recognition techniques are better than the older ones because of their ability to achieve higher detection accuracy levels.

This chapter presented a computational intelligence method for the emotion recognition The previous study based on EEG emotion recognition comprises two steps: data acquisition and processing [22], and the other one is feature extraction and classification [23]. Most studies on data acquisition and processing focus on just a few participants (between 5 and 32). Data were collected using portable EEG systems such as Emotiv EPOC with 14 sensors and clinical EEG systems with different electrodes [24]. Only one study focused on data from 110 individuals, but only 124 sensors are appropriate for the research. Emotional states are aroused by introducing forms of threatening stimuli. Some researchers have found other invoking feelings, such as asking them to conjure memories or to visualize scenarios. While several research studies focused on pictures-based stimuli using the International Affective Picture System (IAPS) [25], there is a lack of an agreed-upon standard for audio and visual stimuli. After having the EEG results, the measurements are preprocessed to eliminate the effects of any unknown objects. This is done by applying band-pass filtering and independent component analysis, which will reduce the impact of artifacts.

The second phase of emotion recognition methods was extracting features from each channel of EEG data. Several varying characteristics could be studied in psychology literature [26]. These features, including classical statistical parameters, autocorrelation coefficients, fractal dimension, non-stationarity index, etc., of time signals can be extracted. Frequency domain features include various frequency bands, power ratios, asymmetrical features, differential entropies, higher-order spectra, spectral moments, and spectral form descriptors [27]. Other time-frequency features are also used in wavelet transformations. These considerations involve functional connectivity measurements such as correlation or phase synchronization between the time signals of the pairs of sensors. The most popular broadcasting bands are the alpha, beta, gamma, and theta bands [28].

Although several methods have been developed for EEG-based emotion recognition, it is essential to choose the right way depending on an individual's physiology. These learning methods are sufficient in today's online education systems. Online learning is necessary when it is challenging to train over the entire dataset due to computational problems. Online learning has been applied successfully in the prediction of sequential emotions.

4.3 Emotion Prediction of EEG Recordings through Computational Intelligence Methods

Let's see how the EEG recordings were used by computational intelligence methods for performing the classification. Figure 4.2 has shown the description of the work discussed in this chapter.

4.3.1 Data Collection

This section explains the experiment conducted to collect and analyze EEG and heart rates to evoke different emotional states from individuals through videos. These datasets with natural vision and hearing were recruited to participate in the experiment. This chapter's discussion is based on four highly accessible auditory-visual databases: HUMAINE [29], MAHNOB-HCI [30], FilmStim [31], and LIRIS-ACCEDE [32]. Therefore, we selected FilmStim for this experiment because it includes film material scenes with similar characteristics and extended duration (40 s to 6 min).

The database consists of 70 movie extracts that neuropsychologists have used for studies of emotional matters. The video archive also contains videos

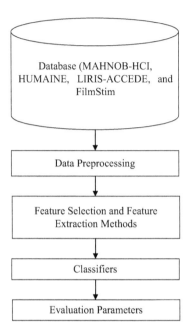

FIGURE 4.2
Description of the work.

about tenderness, amusement, sorrow, rage, disgust, fear, and neutrality. In the experiment, we picked 13 clips from each emotional group (amusement and tenderness). FilmStim database consisted of French-speaking EEG and physiological recordings. HR-EEG and electrophysiological recordings were made in a shielded room using the 256-channel EEG method (EGI, Electrical Geodesics Inc., Eugene, USA). The computational burden was significantly reduced because the initial sampling rate was reduced from 1000 to 250 Hz. The impedance should not be greater than 50 Ω. These computer systems were used for ECG tracking, the patients' respiration, along with the pulse rate and oxygen saturation, as well as oxygen saturation. The chapter is not concerned with such peripheral physiological signals, but they are reported along with the EEG in the study using computational intelligence methods. We also obtained position data from a Sense Wear sensor.

4.3.2 Data Preprocessing

Any of the social variables were incompletely calculated or the mere elicitation of emotions was ineffective (only subjects for whom we succeeded in eliciting emotions for more than 50 percent of the positive videos and more than 50% of the negative videos were retained). In total, 27 individuals participated in the study. We run a simulation and train the algorithm for each emotion. Then, the EEG data were passed through a fifth order Butterworth filter. Based on the visual inspection, the invalid transmission was detected and replaced by interpolated signals from the nearest four channels. Therefore, we used the ICA to eliminate blinks. We discarded components using threshold selection and selected only those with predominantly positive autocorrelation values. However, we ignored the threshold value used because of its unnecessary subject-dependence. As we did not foresee that the subjects would experience extreme emotions during the entire video, we concentrated our study only on each video's 10 s interval, which had the highest effect on emotions. According to subjects, the allocated personality-associated breaks were based on the manual annotation of the material in the videos and the GSR signals' expression. The highest peak value of the GSR may indicate that the individual is highly aroused and may therefore show strong positive or negative emotions. Nevertheless, peak emotional spikes do not accompany scenes with high emotional effects. This is to ensure that the right peaks were chosen for the cycle.

4.3.3 Computational Intelligence with Feature Selection and Feature Extraction

To minimize the number of features for extraction in the source space, there are two methods: one is univariate and the other is multivariate. In this chapter, the discussion is on the univariate feature selection method with the t-test [33]. We have also checked PCA on our data to pick the most useful features, but the results are not as good as those obtained with feature selection, so we have not recorded them.

To compare each condition's brain electrical activity characteristics, we have to decide what brain wave patterns are expected for each situation. The features are derived directly from the signals reported by the EEG sensors in the literature. Sensors carry the signals coming from anywhere on the brain, while not every area can contribute to emotion processing. In this article, we address how to approach category learning by reconstructing the brain's neural network's behavior and by extracting features from the activity. In our system, we use EEG sensors as the sensor space and brain activity of sensor space as the source space features. Brain activity can be restored, and it can be measured by using a large number of dipoles positioned over the cortical surface with an orientation perpendicular to the surface. The EEG characteristics may be correlated with particular brain regions according to the Desikan-Killiany atlas [34].

We have separated the areas into small clusters of dipoles, and so these regions have a similar scale. We are interested in the temporal dynamics of brain signals over time. Denoting by $S \in R^{R \times T}$ i.e. the signal matrix, which characterizes the R brain regions' temporal dynamics at T time points, the measurements $X \in R^{N \times T}$ reported by N sensors on the scalp can be modeled as:

$$X = GS \tag{4.1}$$

Provided a three-dimensional head model, the lead field may be computed numerically, and the lead field may be presumed to be known [35]. The white matter surface mesh is chosen as the source space (each vertex corresponding to a source dipole). The DTI matrix is summarized using brainstorm clustering and distributed using Open MEEG [36]. To recreate the cortical activity within each brain region, it is necessary to solve an inverse problem. This is an ill-posed problem since several more neurons need to be taken into account to obtain a solution to the inverse problem. One specific technique in this field is "WMNE algorithm" [37] that solves an optimization problem.

The sLORETA algorithm as an additional tool was considered in addition to the most promising regularized least-squares algorithms, SISSY and ICA [38]. However, the valence problem findings were slightly worse but possibly still close to those of WMNE. While all R = 549 brain regions contribute to the calculated EEG signals, only part of them is also linked to emotional processes. There has been extensive research in recognizing particular areas in the brain that display positive or negative emotions [39]. We only consider the brain areas that have been found to engage in the development of emotions in functional magnetic resonance imaging (fMRI) studies [39] or intracranial EEG studies [40] for further study and extraction of functions. Features from 274 of the 549 brain regions are derived more precisely. The characteristics are extracted from several 1 s, 2 s, or 5 s segment, segments of the 10 s interval selected for each GSR-based video. All characteristics are measured in the sensor space (for every electrode) and the source space (for each of the 274 emotional source regions). The features used in this paper are described below in Table 4.1.

In Table 4.1, N denotes the number of features extracted for the processing from the sensor space.

TABLE 4.1

Details of Features

Features	Number
SCF	5N
Spectral moments	4N
HOC bands	100N
PSI bands	$N(N - 1)/2$
Band powers	5N
PSI	$5N(N - 1)/2$
HOC	50N

Figure 4.3 shows results of different achievements of metrics for 27 citizens. Whiskers mark the classification spectrum ranging between specificities and sensitivities, averaged over all subjects. The plain-colored bars indicate the results of the sensors, while the hatched-colored bars indicate the results of the sources. The likelihood at which there is one right item is 50%.

4.3.4 Classification

In machine learning world, there are many classification algorithms which can be used for EEG prediction purpose. The most important step in classifying the images through an SVM is to choose the proper functions. The advantage of the linear SVM model is that it needs only one parameter, a desirable value for linearity. Afterward, test results did not affect the training algorithm, so a default value was set. The classification accuracy is assessed based on a cross-validation scheme. For the leave-one-out tests, we used the leave-one-out approach for testing (LVO) [42]. All but one training video is used for the data classification. Classification efficiency is evaluated in terms of precision and sensitivity showing the probability of positive or exact identification. We then measure average specificity and sensitivity ratings from specifics by the district. This hypothesis, despite its high precision, is independent of the probability of each class. The rankings are not based on different class labels. Suppose we are concerned with the outcomes arising from various circumstances. To assess the efficacy of the threshold in various locations, the average difference between specificities and sensitivity values can be used.

4.3.5 Evaluation Parameters

As you can see, multiple texts in each function group illustrated the positive feeling. We will now compare the performance of the classification with various features, from band power to SCF. For most subjects, the similarity of classification scores through characteristics has been observed. It is the consequence that varies from one topic to another. We remember, however, that the best results typically reflect a variety of connectivity characteristics, better order statistics, and spectral momentum from classifications measured over subjects (Figures 4.4 and 4.5). These are typically the lowest output variables when an algorithm of similarity is used. We usually get better results based on connectivity and HOC if we derive characteristics from unique frequency bands rather than from the entire frequency spectrum.

Figure 4.4 shows the EEG data classification with and without ICA-WD preprocessing. The length of the bars reflects how well the question was categorized among all objects, while the inner ends of the bars represent the highest specificity and the outer ends represent the lowest specificity. The green dotted line reflects the chance level classification at exactly 50%.

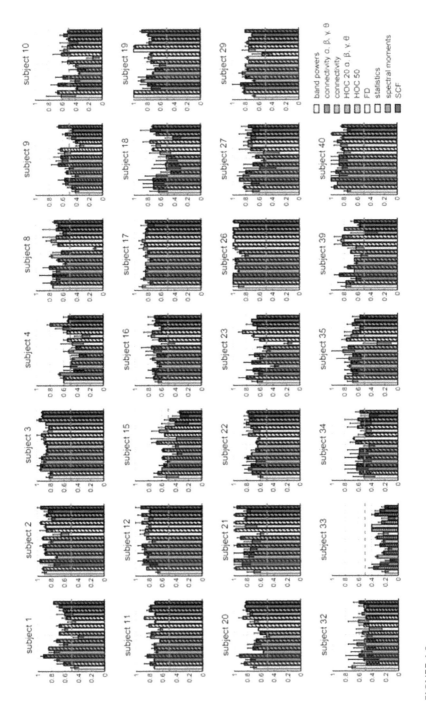

FIGURE 4.3
Comparison performance of different features of the 40 analyzed subjects [41].

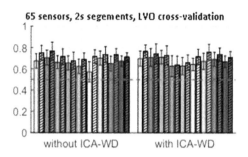

FIGURE 4.4
EEG data classification with and without ICA-WD [41].

In Figure 4.5, the length of the bars reflects how well the question was categorized among all objects, while the inner ends of the bars represent the highest specificity and the outer ends represent the lowest specificity. The parallel lines define where the likelihood lies for a given class. In this graph, there is a modest amount of variability of classification ratings. The red lines mark the 25th percentile, the whiskers mark the 75th percentile, and the colored bar stretches from the 25% percentile to the 75% percentile. In the bottom of the figure, the considered sensors are highlighted in red.

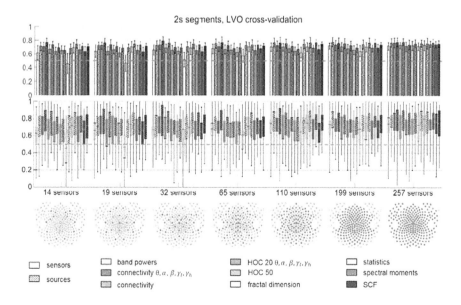

FIGURE 4.5
Classification performance for various electrode configurations [41].

In this chapter, the class distribution for the 27 subjects is examined. These experimental effects are shown in Figure 4.3. The perception of emotion varies considerably between subjects with scores below 40% for subject 30 and scores above 80% for subject 40 for all types of features. We also do a more comprehensive analysis of the five separate datasets for which the highest classification scores have been obtained to clarify the special reasons why the classification results for various subjects are positive variables. These five "best" subjects have their predicted emotions successfully. For the most part, all 12 examples displayed the targeted moods. Topic 40 showed that she was happy/happy with just eight recordings, which was extraordinary compared to other subjects. The brain reported EEG signal was very high for all five participants. Another way to look at these results is that the same images' emotions did not vary. Also, the electrode data for the three groups showed high noise levels. The differences in brain activity can explain the differences in emotional rating performance.

4.4 Conclusion

This chapter presents the computational intelligence in machine learning on a new database of emotion studies that include the HR-EEG and physiological recordings of 40 themes. Then we equate this with seven different types of features from EEG recordings in sensor space and source space. In this way, the processes could be done more robust and stable. There is advanced software designed to interpret facial expressions. The chapter compares the classification efficiency of seven different classifiers extracted from the epileptic patients' electrophysiological records either in the sensor space or in the source space. It can be shown that MRI scans of the brain can improve emotion recognition. The localization stage on average increases the scoring by five percentage points. The incorporation of neural nets and brain activity could yield better classifiers. The price reduction stems to some degree from the increased number of features in the source space. It indicates that their approach to emotion detection can provide a reliable result.

However, ANN can operate with few databases because there is currently no EEG-based emotion recognition technology. In other tests, the precision was greater than 90%. However, this research has certain technical limitations and thus it cannot be possible to generalize to a larger number of participants. The considered number of detectors is dominant in a classification score. The primary reason why classifier preparation needs to be individualized for each subject is due to subject heterogeneity. Our brains can differentiate between positive and negative emotions. Actions such as presenting proof in the courtroom involve excitement or emotional feelings such as fear or anxiety. In [43], Shmuel Bar said that the electrophysiological discriminative features were caused by facial muscles. These results do not prove that state of mind can also be associated with good or poor mood. However, whether muscle function is taken into account or not, brain activity remains an open issue. There are two possible fields of research concerned with the convergence of an EEG signal with data from other physiologic signals. It is useful for the research to combine the valence classification mentioned here with the arousal classification system. By studying, movies with higher levels of arousal would have more a negative impact than movies with positive content. These facts should be reviewed and contrasted so that you can make a decision. ANNs can be used as a promising way to achieve brainwave recognition. Designing models specifically for brain activity with emotions' neural data is specific to adapt deep learning methods.

References

[1] K. Nataliya, "Role of social media in formation and functioning of civil society," *Natl. Acad. Manag. Staff Cult. Arts Her.*, 4, 211–214, 2014.

[2] C. Karyotis, F. Doctor, R. Iqbal, A. James, and V. Chang, "A fuzzy computational model of emotion for cloud based sentiment analysis," *Inf. Sci. (Ny).*, 2018.

[3] R. Nair and A. Bhagat, "Feature selection method to improve the accuracy of classification algorithm," *Int. J. Innov. Technol. Explor. Eng.*, 2019.

[4] P. Ackermann, C. Kohlschein, J. Á. Bitsch, K. Wehrle, and S. Jeschke, "EEG-based automatic emotion recognition: Feature extraction, selection and classification methods," in *2016 IEEE 18th International Conference on e-Health Networking, Applications and Services, Healthcom 2016*, 2016.

[5] T. M. Moerland, J. Broekens, and C. M. Jonker, "Emotion in reinforcement learning agents and robots: A survey," *Mach. Learn.*, 2018.

[6] A. T. Lopes, E. de Aguiar, A. F. De Souza, and T. Oliveira-Santos, "Facial expression recognition with Convolutional neural networks: Coping with few data and the training sample order," *Pattern Recognit.*, 2017.

[7] Z. Haneef, H. S. Levin, J. D. Frost, and E. M. Mizrahi, "Electroencephalography and quantitative electroencephalography in mild traumatic brain injury," *J. Neurotrauma*, 2013.

[8] J. Richiardi, S. Achard, H. Bunke, and D. Van De Ville, "Machine learning with brain graphs: Predictive modeling approaches for functional imaging in systems neuroscience," *IEEE Signal Process. Mag.*, 2013.

[9] J. M. Tuma and J. M. Pratt "Clinical child psychology practice and training: A survey," *J. Clin. Child Psychol.*, 11:27–34, 1982.

[10] A. Craik, Y. He, and J. L. Contreras-Vidal, "Deep learning for electroencephalogram (EEG) classification tasks: A review," *J. Neural Eng.*, 2019.

[11] D. L. Keene, S. Whiting, and E. C. G. Ventureyra, "Electrocorticography," in *Epileptic Disorders*, 2000.

[12] D. P. Subha, P. K. Joseph, R. Acharya, and C. M. Lim, "EEG signal analysis: A survey," *J. Med. Syst.*, 2010.

[13] J. Vorwerk, R. Oostenveld, M. C. Piastra, L. Magyari, and C. H. Wolters, "The FieldTrip-SimBio pipeline for EEG forward solutions," *Biomed. Eng. Online*, 2018.

[14] S. Choi, "Independent component analysis," in *Handbook of Natural Computing*, 2012.

[15] R. Vidal, Y. Ma, and S. S. Sastry, "Principal component analysis," in *Interdisciplinary Applied Mathematics*, 2016.

[16] M. Sugiyama, "Local fisher discriminant analysis for supervised dimensionality reduction," in *ACM International Conference Proceeding Series*, 2006.

[17] B. D. Ripley, "Linear Discriminant Analysis," in *Pattern Recognition and Neural Networks*, 2014.

[18] M. D. Wilson, "Support vector machines," in *Encyclopedia of Ecology, Five-Volume Set*, 2008.

[19] C. Bulac and A. Bulac, "Decision trees," in *Advanced Solutions in Power Systems: HVDC, FACTS, and AI Techniques*, 2016.

[20] D. S. Wilks, *Canonical Correlation Analysis (CCA)*. 2011.

[21] W. Uwents, G. Monfardini, H. Blockeel, M. Gori, and F. Scarselli, "Neural networks for relational learning: An experimental comparison," *Mach. Learn.*, 2011.

[22] A. N. M. M. Yosi, K. A. Sidek, H. S. Yaacob, M. Othman, and A. Z. Jusoh, "Emotion recognition using electroencephalogram signal," *Indones. J. Electr. Eng. Comput. Sci.*, 2019.

[23] M. Hamada, B. B. Zaidan, and A. A. Zaidan, "A systematic review for human eeg brain signals based emotion classification, feature extraction, brain condition, group comparison," *J. Med. Syst.* 2018.

[24] Emotiv Systems, "Brain computer interface & scientific contextual EEG," *EMOTIV EPOC*, 2011.

[25] P. J. Lang, M. M. Bradley, and B. N. Cuthbert, "International affective picture system (IAPS)," *Affect. Ratings Pict. Instr. Man.*, 2005.

[26] A. Abramovitch, G. E. Anholt, A. Cooperman, A. J. L. M. van Balkom, E. J. Giltay, B. W. Penninx, and P. van Oppen, "Body mass index in obsessive-compulsive disorder," *J. Aff. Disord.*, 245: 145–151, 2019. https://doi.org/10.1016/j.jad.2018.10.116.

[27] L. Filipe, F. Fdez-Riverola, N. Costa, and A. Pereira, "wireless body area networks for healthcare applications: Protocol stack review," *Int. J. Distrib. Sens. Networks*. 2015.

[28] J. J. Newson and T. C. Thiagarajan, "EEG frequency bands in psychiatric disorders: A review of resting state studies," *Front. Human Neurosci.*, 2019.

[29] E. Douglas-Cowie et al., "The HUMAINE database: Addressing the collection and annotation of naturalistic and induced emotional data," in *Lecture Notes in Computer Science (including subseries Lecture Notes in Artificial Intelligence and Lecture Notes in Bioinformatics)*, 2007.

[30] M. Soleymani, J. Lichtenauer, T. Pun, and M. Pantic, "A multimodal database for affect recognition and implicit tagging," *IEEE Trans. Affect. Comput.*, 2012.

[31] A. Schaefer, F. Nils, P. Philippot, and X. Sanchez, "Assessing the effectiveness of a large database of emotion-eliciting films: A new tool for emotion researchers," *Cogn. Emot.*, 2010.

[32] Y. Baveye, E. Dellandréa, C. Chamaret, and L. Chen, "LIRIS-ACCEDE: A video database for affective content analysis," *IEEE Trans. Affect. Comput.*, 2015.

[33] A. C. Haury, P. Gestraud, and J. P. Vert, "The influence of feature selection methods on accuracy, stability and interpretability of molecular signatures," *PLoS One*, 2011.

[34] B. Alexander et al., "Desikan-Killiany-Tourville Atlas compatible version of m-CRIB neonatal parcellated whole brain atlas: The m-Crib 2.0," *Front. Neurosci.*, 2019.

[35] A. Gramfort, "Mapping, timing and tracking cortical activations with MEG and EEG: Methods and application to human vision," *Ec. Natl. Super. des Telecommun. - ENST*, 2010.

[36] A. Gramfort, T. Papadopoulo, E. Olivi, and M. Clerc, "Forward field computation with OpenMEEG," *Comput. Intell. Neurosci.*, 2011.

[37] R. D. Pascual-Marqui, "Review of methods for solving the EEG inverse problem," *Int. J. Bioelectromagn.*, 1999.

[38] R. Cannon, C. Kerson, and A. Hampshire, "SLORETA and fMRI detection of medial prefrontal default network anomalies in adult ADHD," *J. Neurother.*, 2011.

[39] K. A. Lindquist, T. D. Wager, H. Kober, E. Bliss-Moreau, and L. F. Barrett, "The brain basis of emotion: A meta-analytic review," *Behavioral Brain Sci.*, 2012.

[40] S. A. Guillory and K. A. Bujarski, "Exploring emotions using invasive methods: Review of 60 years of human intracranial electrophysiology," *Social Cogn. Affective Neurosci.*, 2014.

[41] H. Becker, J. Fleureau, P. Guillotel, F. Wendling, I. Merlet, and L. Albera, "Emotion recognition based on high-resolution EEG recordings and reconstructed Brain sources," *IEEE Trans. Affect. Comput.*, 2020.

[42] T. T. Wong, "Performance evaluation of classification algorithms by k-fold and leave-one-out cross validation," *Pattern Recognit.*, 2015.

[43] M. Soleymani, S. Asghari-Esfeden, Y. Fu, and M. Pantic, "Analysis of EEG signals and facial expressions for continuous emotion detection," *IEEE Trans. Affect. Comput.*, 2016.

5

Finger Vein Feature Extraction Using Contrast Enhancement Dynamic Histogram Equalization for Image Enhancement

R. Anandha Jothi, G. Thenmozhi, J. Nityapriya, and V. Palanisamy
Alagappa University, Karaikudi, India

CONTENTS

5.1   Introduction 63
5.2   Related Works 64
5.3   Proposed Method 65
    5.3.1   Image Enhancement 65
    5.3.2   Normalization 66
5.4   Dynamic Histogram Equalization (DHE) 66
    5.4.1   Dynamic Histogram Equalization Algorithm 66
5.5   Contrast Enhancement Dynamic Histogram Equalization (CEDHE) 67
    5.5.1   Algorithm Steps for CEDHE 67
5.6   Results and Discussion 68
5.7   Performance Evaluation 68
    5.7.1   Quality Measures 68
    5.7.2   Mean Square Error (MSE) 69
    5.7.3   Peak Signal to Noise Ratio (PSNR) 69
5.8   Conclusion 71
Acknowledgment 72
References 72

5.1 Introduction

Most of the person identification systems are based on the biometrics features like physical and behavioral characteristics. Commonly employed physiological biometrics traits are fingerprint, face, iris, hand geometry, ear shape, palm vein, and finger vein. These are related to the shape of the body.

Behavioral characteristics are related to the behavior of the person that involves signature verification technique, speaker recognition technique, and key stroke pattern. Traditional and easily forgettable security systems are replaced by the biometrics characteristics because of its uncountable advantages [1–7]. It is fairly applied to different fields such as medical, financial, retail sale, electronic banking, law enforcement facilities, and other applications where high levels of security or privacy are extremely vital [8].

Finger vein helps to be the most acceptable, widespread biometric characteristics. Finger vein pattern recognition is becoming a promising technology for secured authentication. It is complex to forge the vein patterns because it is placed inside the skin. It does not affect by aging factor and easy to collect [9]. Finger vein-based identification system usually consists of the following stages: (1) image acquisition, (2) pre-processing or image enhancement, (3) feature extraction, and (4) recognition. Image enhancement is an essential step for higher image quality to get better matching performance.

Finger vein images are generally captured using near-infrared light source. Near-infrared light source is passed through the finger to capture the vein images. Hemoglobin in the blood vessel absorbs the infrared light and appears as darker than the surrounding tissues such as muscles and bones. The image captured by this method has some deterioration such as unwanted noise, improper intensity, blurring, and contrast variation. To overcome these problems, image enhancement is required [10]. In this work, histogram equalization is used to enhance the deterioration in contrast.

Dynamic Histogram Equalization is a multi-histogram equalization method that is used to eliminate the power of the higher histogram components on lower histogram components in the image. The main objective of the Contrast Enhancement Dynamic Histogram Equalization is to improve the quality of the image. These two filters are attempted and compared for PSNR and MSE to show the better performance. The rest of the papers contain the following: Section 5.2 describes the related works. Enhancement techniques are presented in Section 5.3. In Section 5.6, Results and Discussion is specified. Finally, we conclude the paper in Section 5.8.

5.2 Related Works

Various authors proposed different histogram equalization techniques for image enhancement. Chen and Ramli et al. studied recursive Mean – Separate Histogram Equalization (RMSHE) to enhance the raw images. In RMSHE, the histogram is split into two pieces of sub-histograms [11]. Sim et al. proposed Recursive Sub-image Histogram Equalization (RSIHE) in which two images were partitioned in DSHIE and the brightness of image is enhanced by multiple level of recursion [12]. Xiao et al. proposed the normalization function to alternate the input image histogram and acquired the specific histogram. The author applied that the probability density function followed by normal distribution function [13].

Sim et al. studied post processing image enhancement technique which includes the process of rescaling the histogram to obtain the output image histogram and this algorithm enhance the range contrast and brightness of the resultant image [14]. Kim et al. studied Bi-Histogram Equalization method that focused on input mean and by using that value, the input image histogram is separated into two histograms and then individually equalizing each part [15]. Sun et al. proposed Dynamic Histogram Equalization that produces the specified histogram vigorously from the input image [16].

5.3 Proposed Method

The main objective of this method is to improve the quality of vein image taken from the database UTFVP. Figure 5.1 depicts the process of image enhancement and feature extraction.

5.3.1 Image Enhancement

Process of enriching the image quality by noise removal, filtering, and contrast adjustment is known as image enhancement. This can be obtained by various techniques.

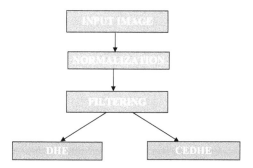

FIGURE 5.1
Schematic diagram of image enhancement.

5.3.2 Normalization

In general, the quantity of data in the image is very large so that the image needs to be normalized into eight-bit grayscale image for increasing the processing speed by the linear transform [17, 18]. This can be done by the following equation (5.1):

$$s(r) = \frac{255}{T_{max} - T_{min}}\left(f(r) - T_{min}\right)T_{min} < f(x) < T_{max} \tag{5.1}$$

where T_{min} represents the maximum gray value and T_{max} represents the minimum gray value. $f(r)$, $s(r)$ are the gray values in the point before and after the normalization of image, respectively.

5.4 Dynamic Histogram Equalization (DHE)

DHE segments the image histograms into sub-histograms until no ruling portion is current in these sub-histograms. It exhibits incessant and improved enhancement of the image than the traditional paradigm [17]. Mean brightness continuation and influences to the intensity diffusion artifacts are managed by DHE. It reinforces the image further than making any obliteration in image essentials. In DHE, enhancement extension can be extending by adjusting the single parameter. DHE is easy to implement and be able to operate in real times because of its transparent and computationally adequate nature.

5.4.1 Dynamic Histogram Equalization Algorithm

Step 1: Acquire the input image and the histogram of the vein image.

Step 2: Determine the local minima in the input finger vein image histogram.

Step 3: Fragment the histograms into sub-histograms based on the local minima in the image to improve the enhancement factor.

Step 4: Assign explicit gray levels to every partition of the histogram.

Step 5: Apply Histogram Equalization on each partition of histogram.

In DHE, local minima is used to divide the histogram. To dispose the insignificant minima, one-dimensional smoothing filter is implemented on the histogram and then it constructs sub-histograms taking the part of histogram that drops between two local minima.

5.5 Contrast Enhancement Dynamic Histogram Equalization (CEDHE)

CEDHE is used to adjust the contrast variations in the vein image and it is the advanced form DHE. The CEDHE is used to control the amount of the gray level region on each partition the greater histogram mechanisms will not completely manage the lower histogram mechanisms. Performing the HE technique straightly on the sub-histograms obtained alone dissent assure quality enhancement in the resultant image. For the minor range sub-histograms, irrelevant contrast enhancement will be created as the sub-histograms tend to be enhanced lesser rather than the others. Precisely implement histogram equalization after reallocating the gray-level allotment outcomes in over-stretching for major range sub-histogram. At last, the traditional HE method was executed in the proposed CEDHE to get the better performance in enhancement.

5.5.1 Algorithm Steps for CEDHE

Step 1: Get the input vein image and its histograms.

Step 2: Normalize the input histogram to safeguard the complete procedure of CEDHE is executed effectively.

CEDHE utilizes the probability density function which can be written as

$$p(r_k) = n_k / n$$

$$k = 0, 1, \ldots, L-1 \tag{5.2}$$

where r_k is the kth gray level and n_k is the number of pixels having the gray level r_k. The total number of pixels in the image is represented as n.

Step 3: Implement the smoothing method on the histogram of given image to safeguard the break point recognition method operates precisely.

Step 4: Perform break point detection histogram on the histogram. There is not static number of breaking points because the number of local minimum appears on each histogram varied.

Step 5: Remapping the gray level allotment to ensure the enhancement of the input image while equalizing directly the sub-histograms can guide to unexpected artifacts within the images.

FIGURE 5.2
DHE enhanced image.

FIGURE 5.3
CEDHE enhanced image.

5.6 Results and Discussion

In this work, we have used UTFVP finger vein database. The database contained with index, middle, and ring vein of both left- and right-hand fingers. The finger vein edges are detected and then subjected with CEDHE and DHE one by one. Figures 5.2 and 5.3 depict the enhanced DHE and CEDHE image. After the image enhancement, the amplification noises are reduced and it automatically corrects variation differences of the original finger vein image.

The quality of the enhanced image looking furnished naturally done by CEDHE compared with DHE image. In this study, we have tested all the images of UTFVP database. The performance of the algorithms was tested with PSNR and MSE. The performance measure of the studied results of the ten right index finger vein images is depicted in Table 5.1. Further, the graphical representation of the performance comparison is shown in Figures 5.2 and 5.3.

5.7 Performance Evaluation

5.7.1 Quality Measures

After the completion of image enhancement, the quality of enhanced vein image has been evaluated by using subsequent measures: MSE, PSNR, and SSIM. These proceedings are universally recognized and used to evaluate the efficacy of the image enhancement [1].

TABLE 5.1

Performance Measures for CEDHE and DHE

Finger Vein Image	DHE		CEDHE	
	MSE	PSNR	MSE	PSNR
001	235	27.13	106	61.51
002	173	28.23	213	60.34
003	373	26.41	271	62.24
004	209	27.28	145	63.14
005	213	26.93	124	61.71
006	263	25.04	114	63.31
007	251	24.43	108	58.80
008	241	22.97	177	59.51
009	231	22.47	155	59.15
010	381	29.44	128	60.7

5.7.2 Mean Square Error (MSE)

The MSE was measured from the collective squared error among original and enhanced image. The MSE is computed by using equation (5.3)

$$MSE = \frac{1}{MN} \sum_{i=1}^{M} \sum_{j=1}^{N} \left[M_1(i,j) - M_2(i,j) \right] 2 \tag{5.3}$$

$M_1(i,j)$ denotes the input image matrix.
$M_2(i,j)$ denotes the filtered image matrix.
$M*N$ indicates the image size.

5.7.3 Peak Signal to Noise Ratio (PSNR)

SNR is a statistical measure of image quality built on the pixel difference among original and resultant finger vein images. The SNR to measure the quality of the rebuilt image equated with the original image, where the value of $S = 255$ only for eight-bit pixel image. The PSNR is fundamentally the SNR while all pixel values are equivalent to the maximum possible value [1]. The value of PSNR was evaluated by the given equation (5.4).

$$PSNR = 10 * log_{10} \frac{R^2}{MSE} \tag{5.4}$$

The proposed enhancement method provides visually cleared patterns after applying CLAHE. The vessel patterns in the vein seem darker that helps to extract the features robustly and easily [20]. We have achieved the desired extraction and accuracy. The performance comparison of the study was clearly shown Figures 5.4–5.6. Figure 5.6 depicts the higher PSNR value

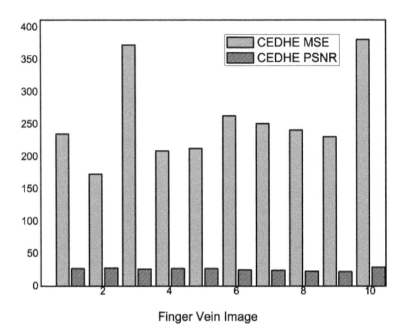

FIGURE 5.4
Performance comparisons of PSNR and MSE for CEDHE.

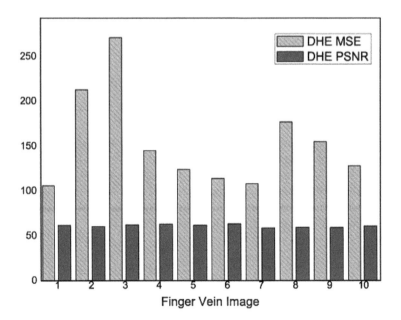

FIGURE 5.5
Performance comparisons of PSNR and MSE for DHE.

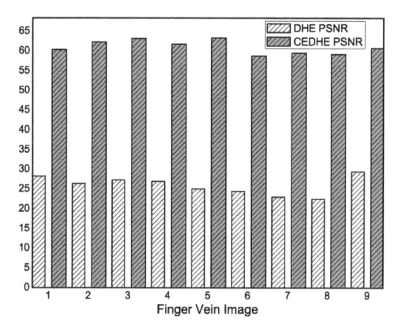

FIGURE 5.6
Performance comparisons of PSNR for DHE and CEDHE.

for CEDHE when compared with DHE-based PSNR. The vein features are extracted with the help of the repeated line tracking algorithm. This algorithm effectively extracted the vein unique patterns.

5.8 Conclusion

We conclude in this chapter that finger vein-based feature extraction and image enhancement are successful by applying DHE and CEDHE techniques. The studied techniques were successfully implemented with MATLAB®. The reconstructed image quality is resolved by measuring the MSE and PSNR. In comparison of equalization techniques, the CEDHE gives higher PSNR and lesser MSE when compared with the DHE technique. The enhanced finger vein images give better recognition accuracy for finger vein identification system.

Acknowledgment

This article has been written with the financial support of RUSA-Phase 2.0 grand sanctioned vide Letter No.F.24-51/2014-U. Policy (TN Multi-Gen). Dept. of Edn. Govt. of India, Dt. 09.10.2018.

References

[1] R. Anandha Jothi, and V. Palanisamy. Performance enhancement of minutiae extraction using frequency and spatial domain filters. *International Journal of Pure and Applied Mathematics* 118.7 (2018): 647–654.

[2] R. Aanandha Jothi, J. Nithyapriya, and V. Palanisamy. *Evaluation of fingerprint minutiae on ridge structure using Gabor and closed hull filters.* In *International Conference on Computational Vision and Bio Inspired Computing* (pp. 663–673). (2018, November). Springer, Cham.

[3] R. Anandha Jothi, and V. Palanisamy. Analysis of fingerprint minutiae extraction and matching an algorithm. *International Journal of Advanced Research trends in Engineering and Technology (IJARTET)* 3.20 (2016): 398–410.

[4] G. Thenmozhi, R. Anandha Jothi, and V. Palanisamy. Comparative analysis of finger vein pattern feature extraction techniques: An overview. *Methods* 6 (2019):40.

[5] G. Thenmozhi, R. Anandha Jothi, and V. Palanisamy. *An Efficient finger vein image enhancement and pattern extraction using clahe and repeated line tracking algorithm.* In *International Conference on Intelligent Computing, Information and Control Systems* (pp. 690–700). (2019, June). Springer, Cham.

[6] R. Anandha Jothi, and V. Palanisamy. A robust and efficient fingerprint minutiae extraction in post-processing algorithm. *International Journal of Biometrics* (Article in press) (2021).

[7] J. Hashimoto. Finger vein authentication technology and its future. In *Symposium on VLSI Circuits, IEEE* (pp. 5–8) (2006).

[8] F. Tagkalakis, D. Vlachakis, V. Megalooikonomou, and A. Skodras, A novel approach to finger vein authentication. *International Symposium on Biomedical Imaging, IEEE* (2017): 659–662.

[9] A. Kumar, and Y. B. Zhou. Human identification using finger images. *IEEE Transactions on Image Process* 21.4 (2012):2228–2244.

[10] E. C. Lee, H. Jung, and D. Kim. New finger biometric method using near infrared imaging. *Sensors* 11.3 (2011):2319–2333.

[11] D. Chen and R. Ramli. Contrast enhancement using recursive mean separate histogram equalization for scalable brightness preservation. *IEEE Trans Consumer Electron* 49.4 (2003):1301–1309.

[12] S. Sim, P. Tso, and Y. Tan. *Recursive sub-image histogram equalization applied to gray scale images.* In *Conference on Pattern Recognition* (pp. 1209–1221) (2007).

[13] B. Xiao, Q. Wang, and X. Zhang. *Image contrast enhancement using normal matching histogram equalization*. In *International Conference on Multimedia Technology (ICMT)* (pp. 1–4) (2010).

[14] S. Sim, Y. Tan, A. Lai, P. Tso, and K. Lim. Reducing scanning electron microscope charging by using exponential contrast stretching technique on post processing images. *Journal of Microscopy* 238 (2010): 44–56.

[15] T. Kim. Contrast enhancement using brightness preserving bi-histogram equalization. *IEEE Transactions on Consumer Electronics* 43 (1997):1–8.

[16] C. Sun, S. Ruan, M. Shie, and T. Pai, Dynamic contrast enhancement based on histogram specification. *IEEE Transactions on Consumer Electronics* 51.4 (2005): 1300–1305.

[17] R. P. Singh, and M. Dixit, Histogram equalization: A strong technique for image enhancement. *International Journal of Signal Processing, Image Processing and Pattern Recognition* 8.8 (2015):345–352.

[18] L. Caixia. The research on finger vein image preprocessing based on mathematical morphology. In *Information Engineering and Applications* (pp. 465–471) (2012). Springer, Switzerland.

[19] W. Z. W. Ismail and K. S. Sim. *Contrast enhancement dynamic histogram equalization for medical image processing application. International Journal of Imaging Systems and Technology* 21.3 (2011):280–289.

6

Song Recommendation Using Computational Techniques Based on Mood Detection

Madhav Kindra, Harshit Garg, Srishti Jhunthra, Vikrant Dixit, and Vedika Gupta
Bharati Vidyapeeth's College of Engineering, New Delhi, India

CONTENTS

6.1 Introduction 75
6.2 Literature Survey 76
6.3 Machine Learning 80
    6.3.1 Support Vector Machine 83
    6.3.2 Naïve Bayes 84
    6.3.3 Random Forest 85
6.4 Deep Learning 85
    6.4.1 Convolutional Neural Network... 86
    6.4.2 Artificial Neural Network (ANN)... 87
    6.4.3 Long Short-Term Memory (LSTM)... 88
6.5 Other Methodologies... 89
6.6 Conclusion 89
References... 90

6.1 Introduction

Music has now been an inseparable part of everyone's life. Music nowadays has been not only a source of entertainment but also one of the best stress relievers. Music and psychology have been deeply interrelated with each other. In a study conducted by USC (University of Southern California) researchers, music affects not only the human brain but also their bodies. The study of human emotions has always been a prominent task in the field of neuroscience and even technology. Humans intend to hide their emotions from the world, but the studies about facial gestures can reveal the person's real mood by analyzing their gestures and facial expressions.

DOI: 10.1201/9781003134138-6

Listening to music can be a source of recreation, but it has many psychological effects too. According to some researches, music can make us healthier, more energetic, and confident; can help in relaxing our mind; and can even help in managing pain. Music has a deep impact on the human brain and body. During sad times, sad music makes us feel relatable to content while joyful music helps us to lift up the mood. A person's mood can be detected by his/her gestures, facial expressions, speech, body language, and many more [1]. Facial detection and speech recognition have been contributing to emotion detection to explore the new areas in the study. Emotion recognition has always been a longstanding goal. The accuracy in detecting the emotion can help in recognizing the extent of the emotion like how sad or happy a person may be.

In this chapter, reviews of various research works that have been published for years have been discussed. Various researchers have worked upon mood prediction and song suggestion via various methods and algorithms. This chapter covers some of the prominent works from the existing works and reviewing their contributions in the field of mood detection and highlighting the gaps in their research works. The chapter is divided as follows: Section 6.2 consists of a detailed review of the prominent research work along with a tabular description. Followed by Sections 6.3 and 6.4 that comprises various methodologies and techniques that have been followed in previous research works, acquiring prominent results has been discussed. The next section includes the overall conclusion along with references.

6.2 Literature Survey

Emotion recognition has always been one of the most interesting and trending topics for years. Over the years, many prominent research works have been published in the field in order to achieve more and more accuracy. Adding music prediction to emotion recognition is one of the trending topics nowadays. In May 2020, BV AR et al. [2] proposed a paper representing a music player application using flutter named "Emotify" which displays a list of recommended songs based on emotion analysis. Their proposed system uses an inbuilt Microsoft emotion recognition API which captures and analyzes emotion from the image through which the system maps the user's emotion with the predefined music playlist. The system provided an accuracy of 60% when tested over one million people. The proposed paper included latency in the starting phase of the application. The application could be further improved by adding some more features such as running a song in a loop, improving the facial recognition process, and focusing more on improving the accuracy for mood detection. Recognizing the emotions was not only limited to still images but also extended to video of a person's face. In 2019, Song S et al. [3] proposed dynamic facial models for video-based dimensional estimation. The authors studied that face is the most important aspect of the prediction of the mood of a user. Thus, Song S et al. used CNN-RNN for their study. The paper concluded that not only the image but also the video of facial expression can be used for detecting a user's emotion. Also, stating that facial shape features generate better results in comparison to facial appearance features. The authors were successfully able to acquire optimal results for facial detection using the video-based dimensional estimation. The proposed paper could be further improved by working on DFM and VSS structure that would increase the accuracy.

Facial recognition was started in the 1970s and still the most studied field in natural emotions machine recognition. In March 2019, Deebika S et al. [4] proposed their study on song prediction using facial expressions. The authors used convolution neural network (CNN) in order to obtain minimal processing and multilayer perceptron. The methodology proposed in this paper is by implementing a facial expression dataset on CNN and mapping the predicted output to the predefined song list. The model obtained the highest accuracy of 85% for an individual emotion. The work could be further improved by analyzing the combinations of different emotions to improve accuracy. At the same time, James HI et al. [5] came up with their study on emotion-based music recommendation systems. The paper uses a linear classifier and support vector machine (SVM) classification technique to carry out the proposed work. The paper reflects accurate results on all the classified emotions except the disgust and fear emotion which reduced the accuracy of the work to 80-85%. The work proposed can be improved by increasing the dataset and focusing on all the emotions. An alternative solution for RPI cameras can also improve the model's performance. CNN is one of the best solutions for image processing and most of the works with high accuracy use CNN for optimal results. In January 2019, Bali V et al. [6] presented their study based on a music player application which displays the recommended song playlist by detecting the emotions of the user. The paper uses CNN and LSTM classification for predicting the song playlist. The paper discusses two methods, i.e., speech and image recognition to predict the song playlist by analyzing emotions of the user through speech or image. The approach used gave the resultant accuracy of 80% to 99% in range. The proposed work can be enhanced in the future by integrating the solution with google play to avoid storage and security issues.

Some works have also been done to create an interactive environment between the user and the interface. In the early 2019, Andjelkovic I et al. [7] proposed a paper on an interactive music player application which takes the user's mood as an input and provides a similar music playlist as the output to the user. The research uses various recommendation systems and effective computing to build an interface for the proposed work. The result acquired from the proposed work improved the user's acceptance and understanding of the recommended song dataset. The results obtained also highlight the huge impact of visual and interactive features in the interface for the music recommendation. The work proposed can be further improved by adding some new features like enabling location and integration of the mood with the location to the interface to make it a user-friendly system. Emotion recognition has not only been limited to images, videos, or speech but has also extended to the sensory response of our body. In 2018, Ayata D et al. [8] gave their analysis on music recommendation based on emotion by using galvanic skin response (GSR) sensors. According to the authors, the GSR sensor generates signals that can be analyzed for emotion recognition. Thus, the methodology proposed in this paper is based on the sensors that the user will have to wear to get his emotions recorded and get the recommended playlist. The research obtained a maximum accuracy of 72.06%. The results obtained have a low accuracy and performance which can be improved with the advancement of the wearable sensor technologies and other hardware equipment used.

Few works in image recognition have been published with the advancements in machine learning. In December 2017, Ramanathan R et al. [9] proposed their work on music recommendation based on emotion recognition by using image capturing and clustering of music. User's image is captured using facial recognition techniques for recognizing the user's emotion. The most suitable music is then recommended based on the predictions made using machine learning models and algorithms. On analysis, the results acquired an accuracy of 72.3%. The results obtained can be further improved by including a number of emotions that could be returned independently irrespective of the number of clusters. Lukose S et al. [10] propose their study on music player based on emotion recognition of voice signals. The study uses machine learning models and algorithms such as SVM, KNN, and GMM for predicting the emotion using voice signals. The overall best accuracy of 81.57% was obtained by SVM for predicting the emotions using voice signals from the user. Nathan KS et al. [11] proposed their analysis in 2017, based on machine learning. The methodology used in the paper is by recognizing speech and image capturing for the prediction of emotions which will further be used for the prediction of preferred song playlist for the user. Different machine learning models were implemented, such as SVM, KNN, and random forest out of which random forest obtained the highest accuracy. The paper also concludes that the neural network performs poorly when implemented on speech recognition. The results could be improved by applying more machine learning models and algorithms. The dataset can also be improved by taking some complex and mixed emotions into consideration.

CNN has been the most accurate and widely used technique not only for emotion detection but also for image processing, object detection, and many

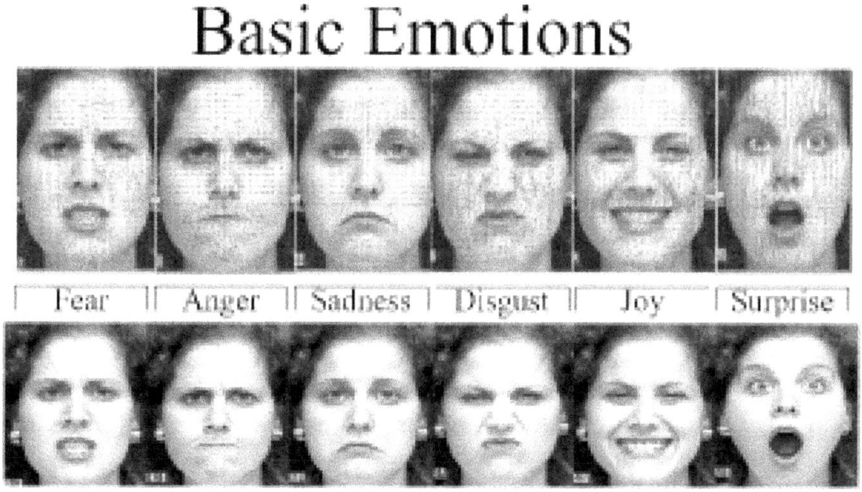

FIGURE 6.1
Basic emotion recognition using deep learning.

more areas. The base of emotion recognition is reading the facial gestures and detecting the mood of the person by categorizing the mood that can be analyzed using deep learning. Figure 6.1 shows the basic categories of moods that most of the works used to categorize their dataset.

Deep learning and neural networks have been the major contributors in emotion recognition and music prediction. Some prominent works on the same have been published in the recent years. In 2017, Gilda S et al. [12] presented their study on an effective cross-platform music player, which will predict the facial gestures of the user and recommend the corresponding music playlist by using deep learning. The study completely focuses on artificial neural networks and multi-layered neural networks. The accuracy of the results obtained after applying deep learning concepts was 90.23% after classifying four major emotions. In future, the results for more emotions can also be recorded and analyzed to check the performance of the model. The model can also be enhanced by adding a variety of songs of different languages and emotions to increase the accuracy. In the same year, Lopes AT et al. [13] proposed their work on facial emotion recognition using CNN. The study is based on the emotion detection using CNN by using facial gestures of the user. The study performed uses CNN and obtained an accuracy of 96.76% on the dataset provided. The study describes the methodology used for obtaining an optimal solution for the prediction of emotions using CNNs. The study acquired proficient results and further can be considered for integrating with different areas using emotion recognition by facial gestures. In 2016, Kamble SG et al. [14] proposed their study on facial expression-based music players. The study proposed uses the concept of Euclidean distance classifier for predicting the music playlist based on the facial expressions. The facial expressions are analyzed using principal component analysis (PCA) and Euclidean distance classifier using an inbuilt camera to cut off the designing cost of the model. The results acquired an accuracy of 84.82% in recognizing the correct facial expression and displaying the corresponding music playlist. The work can be improved by making the model compatible with analyzing mixed emotions using facial gestures. Khorrami P et al. [15] in 2016 proposed their study to improve the results of emotion recognition from video data using deep neural networks. Different types of neural networks analyze how much each neural network component contributes to the system's overall performance. The study concluded that a single frame CNN outperforms all the combinations of neural networks even after applying adjusting hyperparameters. The study also concludes that when an extensive hyperparameter is taken with CNN+RNN combination, the model performs and gives accurate results. The study can be improved by applying the models on a huge dataset to obtain more accurate and precise results for the emotion recognition using video dataset.

In 2018, Wang J et al. [16] proposed a research on the facial recognition based on CNN. The paper proposes the actual working of facial recognition using CNN. The paper discusses the major aspects and the use of CNN to obtain an optimal and accurate result for the facial detection. The study can be used in implementing CNN for facial recognition which could further be used for analyzing the emotion of the user. The paper concludes that CNN is the most appropriate model for implementing and predicting facial recognition model. The prominent works discussed in this section are tabulated in Table 6.1.

Thus, music and emotion have always been interrelated to each other from medieval times. As noted above, many researches have been accomplished and various methodologies have been proposed over time to achieve success in various domains of the project. The next section consists of some of the most common methodologies that attained high accuracy and are even currently being implemented with further modifications to overcome the gaps existing in previous implementations.

6.3 Machine Learning

Machine learning is a concept of training a model, building an algorithm, and improving these results with experience. In simpler words, machine learning is training a model over a dataset and testing it to predict future outputs. It is

TABLE 6.1

Comparative Analysis of Different Studies for Music Recommendation Based on Emotion Recognition

Author	Problem Addressed	Basic Approach	Achievements	Research Gaps
BV AR et al. (2020)	To generate songs based on emotion analysis.	Flutter-based application along with Microsoft API.	Accuracy: 60%	1. Adding more features. 2. Improving accuracy.
Song S et al. (2019)	To study video-based dimensional estimation for mood prediction.	CNN-RNN-based models.	The model obtained optimal results.	1. Working on DFM structure. 2. Working with VSS structure to improve accuracy.
Deebika S et al. (2019)	To determine music by analyzing user's expressions.	CNN (convolution neural network)-based models and algorithms.	Accuracy: 85%	1. Analyzing the combination of different emotions. 2. To improve accuracy.
James HI et al. (2019)	To study different classification techniques.	Linear and SVM classifier to predict the results.	Accuracy varies in the range of 80% to 85%.	1. To find an alternative solution for RPI cameras. 2. Implementing more models and classifiers to improve accuracy.
Bali V et al. (2019)	To build a real-time application for users.	Usage of CNN and LSTM classifier for different predictions and model generation.	Accuracy varies in the range of 80% to 99%.	1. By integrating the solution with google play. 2. To avoid security and storage issues.
Andjelkovic I et al. (2019)	To find user's mood specified song playlist.	Build a user-friendly interface for the users.	Highlights the huge impact of visual and interactive features.	1. Adding more features to the system. 2. Integrating mood with location.
Ayata D et al. (2018)	To study sensors technology for detecting moods and expressions.	Applying GSR sensors for predicting facial emotions and gestures.	Accuracy: 72.06%	1. Improving the accuracy. 2. Enhancing the hardware components to improve efficiency.
Ramanathan R et al. (2017)	To recognize emotions by focusing on image capturing and clustering techniques.	Image recognition and machine learning models.	Accuracy: 72.3%	1. Including a greater number of emotions.

(Continued)

TABLE 6.1 (*Continued*)

Comparative Analysis of Different Studies for Music Recommendation Based on Emotion Recognition

Author	Problem Addressed	Basic Approach	Achievements	Research Gaps
Lukose S et al. (2017)	To generate a music system based on voice signals.	Different machine learning models and algorithms.	Accuracy: 81.57%	1. Improve accuracy. 2. Apply different models to build a new algorithm.
Nathan KS et al. (2017)	To recognize speech and images for mood prediction.	Using different machine learning models.	Out of all the models, random forest has highest accuracy and precision.	1. Improving dataset to analyze some complex situations.
Gilda S et al. (2017)	To build a smart and effective cross-platform music player.	Artificial neural network and multi-layered neural network.	Accuracy: 90.23%	1. Addition of a variety of songs of different language. 2. Testing of model on a different dataset.
Lopes AT et al. (2017)	To study convolution neural network for generating mood prediction models.	Performing CNN for recognition of facial expressions.	Accuracy: 96.76%	1. Integration of the models with different areas.
Kamble SG et al. (2016)	To introduce Euclidean distance classifier for music prediction based on facial expressions and gestures of a user.	Introducing PCA and Euclidean distance classifier for detection.	Accuracy: 84.82%	1. Making model more compatible to analyze mixed emotions. 2. Improving accuracy.
Khorrami P et al. (2016)	To recognize video data for emotion recognition.	Using deep neural network for recognition and song predictions using video analysis.	Introduced a new extensive parameter which performs and gives accurate results.	1. Applying model on a huge dataset. 2. Increase precision using video-specific dataset.
Wang J et al. (2018)	To solve the music recommendation problem using an effective approach.	Building CNN and deep learning-based models.	Concludes that CNN has a major aspect in emotion recognition.	1. By obtaining an improved model and algorithm to improve accuracy. 2. Calculate more mathematical results.

a subset of artificial intelligence that ensures the model's accuracy and efficient working. Machine learning includes three types of learning concepts:

1. Supervised learning;
2. Unsupervised learning;
3. Reinforcement learning.

Out of these learning techniques, supervised learning has always been in trend in most of the researches for predicting emotions based on the user's mood [17]. Supervised learning models use the mapping of functions to make predictions by training the machine and then testing it on various inputs. Some of the most frequently used techniques used in previous research works are as follows.

6.3.1 Support Vector Machine

SVM algorithm is one of the supervised learning models, which is associated with a learning algorithm that analyzes data used for classification and regression analysis. This algorithm is mainly used to solve two-group classification problems by defining a decision boundary for the two groups.

Figure 6.2 represents the best hyperplane for distinguishing between two class types. Various hyperplanes can be made for a given dataset but, the best suited is the one that has a huge margin between the two classes. Therefore,

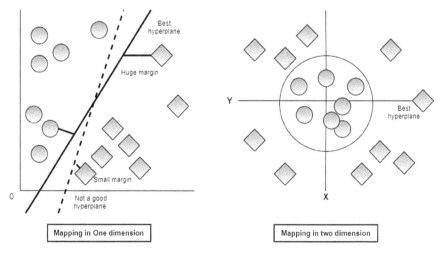

FIGURE 6.2
Mapping of hyperplane in SVM.

SVM can be considered for comparing the data inputted, i.e., human face, and categorize which emotion matches the input as SVM helps in distinguishing between multi-class datasets [5, 10, 11].

6.3.2 Naïve Bayes

Naïve Bayes is another supervised learning model that is based on the concept of Bayes theorem. Two assumptions are made in this model, one is that all predictors are independent and the other is that predictors have equal weights in the resulting output.

Figure 6.3 shows the working procedure of naïve Bayes classifier. Naïve Bayes model is trained and then a prediction is made for other instances. Among various machine learning models, naive Bayes can be selected to provide better results as it is based on Bayes theorem due to which it can be considered [11].

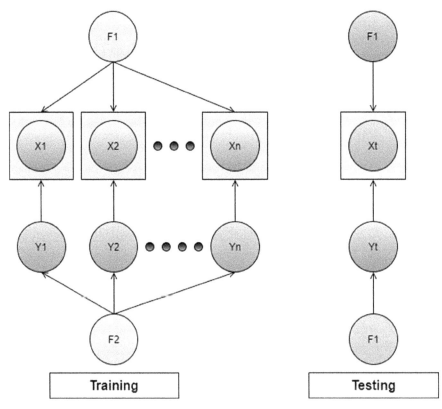

FIGURE 6.3
Naïve Bayes training and testing procedure.

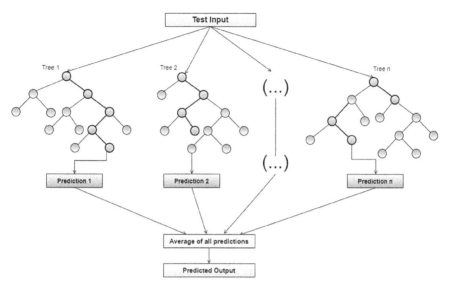

FIGURE 6.4
Random forest procedure.

6.3.3 Random Forest

Random forest also known as decision forest is a machine learning algorithm that uses decision trees for predicting the results. Random forest outperforms other machine learning models and few ensembles most of the time because of its key concept which consists of numerous unrelated trees that produce ensemble predictions that are more accurate than other models.

As shown in Figure 6.4, the output predicted in random forest is determined by taking the average of all the individual predictions made by every decision tree. On implementing various machine learning and even ensembles, random forest outperforms other machine learning models with the lowest error in predictions [11].

6.4 Deep Learning

Deep learning is a subset of machine learning. It can be categorized as the special category of machine learning in which neural networks are trained to learn large amount of dataset. Due to its advantage to build an adaptive environment over huge dataset, this learning technique is used for detection purposes. Facial detection, object detection, and image processing can be made easy and accurate using deep learning [18]. Deep learning involves building of neural networks to test and train the given data. There are various types of deep learning concepts that can be used for mood prediction using facial expressions and gestures.

6.4.1 Convolutional Neural Network

CNN comes under the category of deep learning technique which is used for recognizing, analyzing, and visualizing different images; based on the dataset, the neural network is trained [3]. This neural network is also termed as space invariant artificial neural network due to its formation. CNN is based on multilayering process where the output of lower layer serves as an input to the higher layer of the neural network. The layers are fully connected with neurons following a hierarchical approach to compute data using simple techniques.

The architecture of CNN is an analogous connectivity like the patterns of neurons in the human brain. This neural network was inspired by the visual cortex. As shown in Figure 6.5, this neural network consists of various layers which are fully connected with each other. Therefore, CNN can be considered as the most beneficial technique for emotion detection using the facial expression dataset as an input. The neural network can be trained using the facial expressions and gestures dataset and a mood corresponding to that dataset can be predicted using the power of CNN.

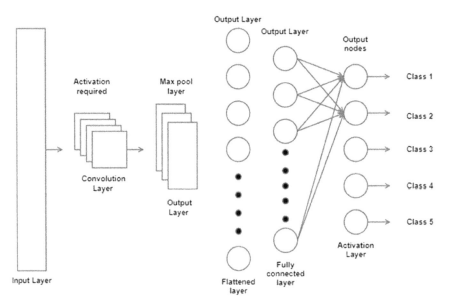

FIGURE 6.5
CNN architecture.

6.4.2 Artificial Neural Network (ANN)

Neural networks are considered the most powerful technique when it comes to massive dataset. Thus, artificial neural network is one of the most powerful and the most popular model. Though, this neural network was not able to show an outstanding performance in its early stage. But, in the recent times, this neural network has gained a lot of popularity because the results acquired by this model cannot be competed with those by another model or neural network. ANN is basically an information processing model that is based on the concept of a biological nervous system. These neural networks are considered as loosely modeled or have a process consisting of nonlinear relationships between inputs and outputs that are being executed parallelly.

ANN has the ability to learn a huge dataset quickly. The information or dataset trained flows through the neural networks quickly which makes this neural network strong and powerful which is considered beneficial to perform a variety of tasks.

As shown in Figure 6.6, the ANN architecture states the following points:

1. The architecture consists of three layers, i.e., the input layer, hidden layer, and the output layer. This is also known as the MLP (multilayer perceptron) due to the occurrence of multiple layers.

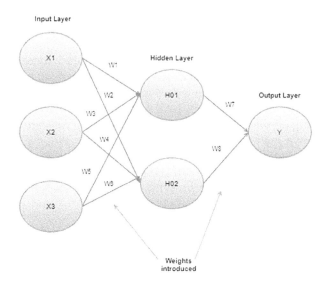

FIGURE 6.6
ANN architecture.

2. The hidden layer can include more than one layer. These layers can be seen as the distillation layers where the patterns obtained from the input can be distilled and then passed on to the next layer. This makes the network faster and accurate as it helps to identify the important information at each consecutive layer excluding the redundant information.

3. The activation function of the neural network includes two key points:

   a. This helps to convert the input given to an accurate and powerful output.

   b. It works on the non-linear relationships.

4. The hidden layers act as the deciding factor to make the final prediction obtained at the output layer.

5. Weights carry the maximum importance which are associated with the inputs to obtain accuracy results.

Thereby, artificial neural network can be used for image processing, face detection, and the mood corresponding to the face detected. The neural network can be trained using the expression dataset with the corresponding moods. After the processing of the dataset, the artificial neural network will be able to process and predict the emotion corresponding to the input provided.

6.4.3 Long Short-Term Memory (LSTM)

In the past years, when neural networks were introduced, there also aroused a problem "the short-term memory." All the recurrent neural networks faced this short-term memory problem which made it difficult to carry huge information in a short amount of time. To solve this problem LSTM, it processes data by passing on the information and propagating it forward.

LSTM follows the concept of the cell state with various gates. The cell state acts as a highway bridge that transfers the relative information through a sequential chain. The cell states hold a special feature that it can carry all the relevant information while processing the chain. Thus, the information stored in the initial steps is stored till the end states, which reduces the effects of short-term memory. Figure 6.7 shows the working of LSTM model where the input, output, and process are based on cell state and gate methodology. This concept can be used as a classification method for predicting emotions of a user using his/her facial expression and gestures. This can be implemented by taking a training dataset as an input and processing the dataset till the desired output, i.e., predicted emotion is obtained.

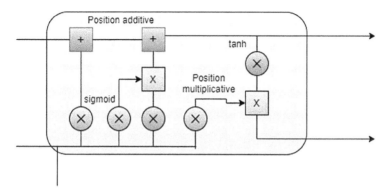

FIGURE 6.7
Working of LSTM.

6.5 Other Methodologies

Other than machine learning and deep learning concepts, various innovative methodologies have also been used to recommend song based on user's mood using facial expression and gestures. PCA [14] is one such technique that uses an orthogonal transformation that builds a set of non-related variables from the correlated variables. This concept has been widely used for data analysis and in various predictive machine learning models. Haar cascade [9] has also been used in face detection to extract features rapidly. The advantage of using this technique is increased speed and it takes multiple images of a single person for better recognition. Other than using models and algorithms, flutter [2] has also contributed to the field of image recognition and face detection. It consists of a Firebase ML kit which includes various APIs for the implementation of image recognition. Despite using algorithms, hardware technologies such as GSR sensors [8] are used for measuring the electrical conductance of the skin. In response, human emotion is recognized by computing sweat gland activity.

The next section consists of description of the previous research work that has been performed in various domains for song recommendation based on emotion recognition using facial expressions. The section covers the methodology used, along with their success and research gaps in their respective studies.

6.6 Conclusion

In this study, we have come across various researches performed on music recommendation using facial expressions and gestures using different techniques and datasets such as video-based, image-based, speech-based recognition, and many more. Music has become a soul part of our life. A person's mood can be modified using different kinds of music at the same time. Music can be treated as the psychological part of the human brain where the tone, pitch, lyrics combine together to make a person to relate his social and personal issues with it. With the evolution of the modern era, music has also been evolved. In the past times, a bunch of sticks, utensils, etc. was used to create music of different notes. But as the evolution took place, this equipment was replaced by modern instruments which can be played to create different emotions of music. Being attached to the emotions music produce, it is important for analyzing a user's mood to help the user recommend a playlist of his desired mood. However, music acts as a source of entertainment but can also be considered as a utilitarian for the users to come across a system-predicted playlist instead of searching and scrolling different applications manually. Considering the analysis made in this study, one can work on improving the drawbacks and disabilities in the researches to come up with a new and innovative concept for prediction. Song recommendation using facial expression and gestures is therefore necessary and has been performed by many researchers for the past years. These researchers therefore have been recorded and also tabulated in Table 6.1.

References

[1] Gupta V, Juyal S, Singh GP, Killa C, Gupta N. Emotion recognition of audio/speech data using deep learning approaches. *Journal of Information and Optimization Sciences.* 2020;41(6):1309–1317

[2] Akshobhya Rao BV, Asokan FR, Firdous H, Prerana GP, Shyam GK. Emotion based music player (emotify). *International Journal of Advanced Research in Computer Science.* 2020;11(Special Issue 1):17.

[3] Song S, Sánchez-Lozano E, Kumar Tellamekala M, Shen L, Johnston A, Valstar M. *Dynamic facial models for video-based dimensional affect estimation.* In *Proceedings of the IEEE International Conference on Computer Vision Workshops* 2019 (pp. 1608–1617).

[4] Deebika S, Indira KA. *A machine learning based music player by detecting emotions.* In *2019 Fifth International Conference on Science Technology Engineering and Mathematics (ICONSTEM)* 2019 March 14 (Vol. 1, pp. 196–200). IEEE.

[5] James HI, Arnold JJ, Ruban JM, Tamilarasan M, Saranya R. Emotion based music recommendation system. *Emotion.* 2019;6(3).

[6] Bali V, Haval S, Patil S, Priyambiga R. Emotion Based Music Player. 2019 emotion.;9:8.

[7] Andjelkovic I, Parra D, O'Donovan J. Moodplay: Interactive music recommendation based on Artists' mood similarity. *International Journal of Human-Computer Studies.* 2019;121:142–159.

[8] Ayata D, Yaslan Y, Kamasak ME. Emotion based music recommendation system using wearable physiological sensors. *IEEE Transactions on Consumer Electronics.* 2018;64(2):196–203.

[9] Ramanathan R, Kumaran R, Rohan RR, Gupta R, Prabhu V. An *Intelligent Music Player Based on Emotion Recognition.* In *2017 2nd International Conference on Computational Systems and Information Technology for Sustainable Solution (CSITSS)* 2017 December 21 (pp. 1–5). IEEE.

[10] Lukose S, Upadhya SS. *Music player based on emotion recognition of voice signals.* In *2017 International Conference on Intelligent Computing, Instrumentation and Control Technologies (ICICICT)* 2017 Jul 6 (pp. 1751–1754). IEEE.

[11] Nathan KS, Arun M, Kannan MS. *EMOSIC—An emotion based music player for Android.* In *2017 IEEE International Symposium on Signal Processing and Information Technology (ISSPIT)* 2017 December 18 (pp. 371–376). IEEE.

[12] Gilda S, Zafar H, Soni C, Waghurdekar K. *Smart music player integrating facial emotion recognition and music mood recommendation.* In *2017 International Conference on Wireless Communications, Signal Processing and Networking (WiSPNET)* 2017 March 22 (pp. 154–158). IEEE.

[13] Lopes AT, de Aguiar E, De Souza AF, Oliveira-Santos T. Facial expression recognition with convolutional neural networks: coping with few data and the training sample order. *Pattern Recognition* 2017;61:610–628.

[14] Kamble SG, Kulkarni AH. *Facial expression based music player.* In *2016 International Conference on Advances in Computing, Communications and Informatics (ICACCI)* 2016 September 21 (pp. 561–566). IEEE.

[15] Khorrami P, Le Paine T, Brady K, Dagli C, Huang TS. *How deep neural networks can improve emotion recognition on video data.* In *2016 IEEE international conference on image processing (ICIP)* 2016 September 25 (pp. 619–623). IEEE.

[16] Wang J, Li Z. *Research on face recognition based on CNN.* In *Proceedings of the IOP Conference* 2018 July (pp. 170–177).

[17] Gupta V, Singh VK, Mukhija P, Ghose U. Aspect-based sentiment analysis of mobile reviews. *Journal of Intelligent & Fuzzy Systems.* 2019;36(5):4721–4730.

[18] Jain N, Gupta V, Shubham S, Madan A, Chaudhary A, Santosh KC. Understanding cartoon emotion using integrated deep neural network on large dataset. *Neural Computing and Applications.* 2021;21:1–21.

7

Deep Learning Classification of Retinal Images for the Early Detection of Diabetic Retinopathy Disease

N. Indumathi, B. Kalanjiyam, and R. Ramalakshmi

Kalasalingam Academy of Research and Education, Krishnankoil, India

CONTENTS

7.1 Introduction 93
7.2 Classification of the Diabetic Retinopathy 97
7.3 Related Study 99
7.4 Methodology 101
    7.4.1 Dataset 101
    7.4.2 Instruction for the Grading 102
    7.4.3 Development of the Algorithm 103
    7.4.4 Purpose of CNN 103
    7.4.5 EfficientNet 104
    7.4.6 Evolution of Algorithm 105
    7.4.7 Performance Analysis 106
7.5 Result and Discussion 107
    7.5.1 Data Analysis 107
    7.5.2 Convolutional Neural Network 108
    7.5.3 EfficientNet Model 108
    7.5.4 Model Prediction 109
    7.5.5 Accuracy 110
7.6 Conclusion 111
Acknowledgement 111
References 111

7.1 Introduction

Deep learning is a class of computational approaches that enable an algorithm to programme itself through understanding from a huge set of examples that exhibit the desired behaviour, eliminating the need to explicitly describe rules. The use of these approaches to medical imaging necessitates additional testing and validation. In recent decades, a new computer intelligent system that uses machine learning methodologies has recently been built and is currently in use in many different fields, specifically designed for implementation in the field of medical to diagnostic systems of health complications. This learning method is used to identify the significant aspects of the biomedical dataset and use those features to produce predictions of new facts about the existing data. Machine learning has started to incorporate 2D and 3D images.

Diabetes is one of the world's most insidious epidemics. In particular, the Indian subcontinent has the highest prevalence of diabetes. This has resulted in the emergence of new diseases such as diabetic retinopathy (DR). It is critical to developing an automated analysis model to increase the ophthalmologist's job scope and reduce patient suffering. More than 422 million people worldwide have diabetes, with India ranked third worldwide for having the highest numbers of diabetes cases among the three countries listed. An increase in the number of those who reside in the US, China, Brazil, India, and Indonesia are accompanied by an expansion of the global population, as 42% of the world's population is estimated to now reside in these four countries [1]. Next to cataracts, glaucoma is the world's only other largely-asymtomatic retinal disease. This causes the retinal injury and subsequent loss of elasticity, causing the retinal visualization to inflate. An analysis of a fundus image from 2D can be difficult. Avoiding patients' eyesight loss is imperative to choosing the correct course of action. Predominantly, the existing situation does not now have an efficient method. It only took a short period of time for multiple studies to verify that the fundus picture of the retina could be uncovered by different image analyses [2]. A major portion of the increased global number of people who have diabetes can be attributed to the *Lancet* study, which discovered diabetes in the Indian and Chinese populations and the fact that these data were released in the US. Diabetes increases the risk of several medical concerns, one of which is the possibility of having diminished eyesight as a result of retinopathy of prematurity [3]. The retina is an optic nerve tissue photosensitive layer that lies inside the eyeball. A variety of diseases can cause retinal damage, which can lead to irreversible vision loss. The population ageing has emerged to be a major demographic trend globally and is expected to grow in patients with chorio-retinal diseases like age-related macular degeneration (AMD) and diabetic mellitus retinopathy (DMR) [4]. Blindness may result from AMD. In diabetes mellitus (DM) patients, the DMR is a common lifestyle disease. Macular degeneration (MD), often known as age-related macular degeneration, afflicts the elderly. Figure 7.1 represents the vision loss due to the macular degeneration [5]. Currently, the fastest-growing demographic in industrialized countries is people aged 65 years and older; this growth in AMD risk and effects will only increase in the future. As a result, there is a great need to develop preventative measures to combat AMD and to halt the onset of vision loss at an early stage. Figure 7.2 illustrates the variance of age-related macular degeneration [6]. DR is the most prevalent eye condition caused by high blood sugar levels in the human body, which destroys the retina's blood vessels (DR). Early retinopathy diabetes can produce vision problems or possibly create no symptoms at all [7]. An estimated 422 million diabetes patients worldwide are estimated by the World Health Organization (WHO). The prevalence of diabetes has increased from 4.7 to 8.5% worldwide. The WHO estimates that in 2030, there will be a reported growth in the number of people with diabetes from 171 million to an additional 366 million [8]. The number of DR patients is expected to grow to 191.0 million by 2030, given the number of diabetic patients each day. Presently, the prevalence of DR and

FIGURE 7.1
Macular degeneration (MD).

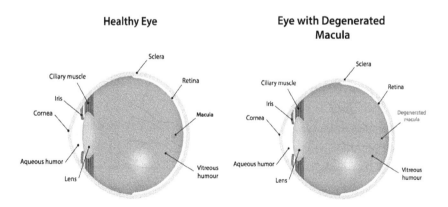

FIGURE 7.2
Age-related macular degeneration (AMD).

macular oedema (DME) is 35% and 7%. The result of diabetes and its complications is considerable financial losses for patients and their families [9]. DM was once a primary problem for India's urban population. DR, also known as diabetic eye disease, occurs when the retina is damaged as a result of diabetes, according to new research. It is a systemic condition that affects 80% of diabetic individuals who have been diagnosed for more than 20 years [10]. Recent studies indicate that 90% of new cases of eye disease may be averted if thorough and careful monitoring and treatment of the eyes were employed. DR is increasingly common as the duration of diabetes increases [11]. The severity of clinical DR is divided into five levels. Although many patients do not suffer from clinically detectable DR early after DM diagnosis, the retinal blood flow, leukocytes, density of the cell membranes, and loss of retinal pericytes are known to change structures and have physiologic changes. An abnormality is commonly used in the presence and size for determining a disease's severity. A critical step in the diagnostic procedure is the detection of conditions such as venous beads, microaneurysms, and haemorrhage. Microaneurysms are 100–120 m blood clots that usually have a circle-like shape. Blood leakage is an irregular development of miniature blood vessels. Haemorrhaging is the leakage of blood from the damaged blood vessels. The vein beading shows, therefore, how the veins placed next to the arterioles occluded fundamentally expand. Non-proliferative DR and proliferative DR are two different kinds of DR [12].

To avoid vision loss, preventative care was required. The discovery of exudates in the fundus view, as shown in Figure 7.3, allowed for the diagnosis of DR. There are numerous feature improvement approaches, such as matched filtering, thresholding, or optimum wavelet transform, as well as neural networks, K nearest neighbour (k-NN) classifiers, AdaBoost, Naïve Bayes, random forest, and Support Vector Machine (SVM), used ensemble machine learning techniques to predict DR classification and which are all employed in order to quantify the severity of DR [13].

I aimed to train a convolutional neural network (CNN), a popular artificial intelligence (AI) technique, to identify fundus images, which it could place into one of five categories:

- There is no DR present;
- DR that is not severe;
- DR that is moderate;
- DR that is severe;
- DR that is proliferating.

Retinopathy is diagnosed based on the patterns of disease that appear in the different stages of the disease. Train a CNN on the photos, and estimate the accuracy model for the images using EfficientNet.

(a) normal vision (b) vision affected by advanced DR

(c) blurred vision due to DME (d) diabetic-affected affected vision

FIGURE 7.3
Vision afflicted by a variety of illnesses. (a) Normal vision. (b) Vision affected by advanced DR. (c) Blurred vision due to DME. (d) Diabetic-affected vision.

7.2 Classification of the Diabetic Retinopathy

The screening for DR includes an important component known as retinal image quality assessment (RIQA). In diabetes, the major cause of vision loss is DR. Screening to detect the disease in its early stages is crucial to providing therapy as soon as possible. Additionally, medical photographs collected and analyzed in this manner consistently fail to correctly diagnose the problem, and the work of ophthalmologists is thus wasted [14]. DR is the most commonly seen consequence of diabetes. People with type 1 or type 2 diabetes are commonly affected by this condition. Diabetes requires regular eye exams because of the slow rate of disease progression. DR prediction is

FIGURE 7.4
Stages of diabetic retinopathy.

a technique in which a patient presents a series of photographs of his or her retina and then the doctor analyzes these photos to see if the patient has DR. DR grading is the process of identifying the point in time when DR is first applied to the input photos. A vast corpus of photos is used in this method. Each image has a set of values ranging from 0 to 4 [15]. According to disease severity, the NPDR is classified into stages. Several stages of fundus image development based on the severity, as shown in the example DR pictures in Figure 7.4. Table 7.1 explained about the structure of DR.

A complication induced by high blood sugar levels that affects the blood vessels in the light-sensitive tissue (retina) in the back of the eye, resulting in the blood vessels haemorrhaging and distorting vision. It is the most frequent cause of vision loss in people with diabetes, and it is the most significant cause of blindness or vision impairment in adults [16]. Access to the

TABLE 7.1

Representation of the Structure of Diabetic Retinopathy

Levels	Stages	Identification
Class 0	No DR	Normal
Class 1	Mild DR	In the early stages of retinopathy, microaneurysms form in the retina's blood vessels. The little aneurysm could cause a fluid leak into the retina.
Class 2	Moderate DR	The second stage of retinopathy can cause blood vessels that supply the retina to enlarge and distort. Reduced blood flow is one potential side effect of this treatment. In both cases, the alterations in the retina can result in DME.
Class 3	Severe DR	The development of retinopathy from stage 2 to stage 3 results in an increase in the restriction of blood flow to the retinal area that produces growth factors, indicating to the retina to begin generating new blood vessels.
Class 4	Proliferative DR	During the fourth and most advanced stage of retinopathy, growth factors generated by the retina activate the creation of new blood vessels that extend along the internal surface of the retina and into the vitreous gel (fluid that fills the eye). In part because these new blood vessels are more weak, they are more prone to leakage and bleeding.

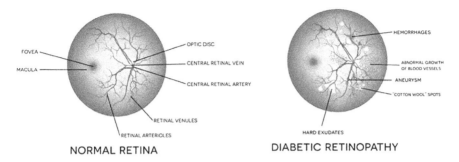

FIGURE 7.5
Image representation of normal and DR.

health information system's data may threaten the security of patients' medical photos, thereby compromising the integrity of the health system. There are numerous studies currently being undertaken to improve the security of medical imaging, and the overall purpose of the current research is to propose an encryption strategy that incorporates chaos theory [17].

Figure 7.5 represents the differentiation between normal vision and DR vision.

7.3 Related Study

Researchers in DR have done a great deal of work depending on their areas of research and interest. Researchers have developed and executed different ways of machine learning, as shown by the linked research in the medical area, although the study comparison among these ways of profound learning remains lacking in diabetes retinopathy. The outcomes and findings of various machine learning algorithms for DR are thereby proven to be a revolutionary technique.

Shankar et al state that the classification is an important part of many different areas, such as health care. SVM classification models as well as its group are utilized to increase the performance of the classifier by means of better grey wolf optimization [18]. Raman et al. are dedicated to establishing a computer-aided detection system that identifies faulty retinal imagery, while recognizing abnormal characteristics of retinal fundus pictures. The methodology proposed in this study focuses on improved images, noise filtration, vascular diagnosis, and the identification of the optic fluid and then extracts features and classifies various stages of diabetes retinopathy as mild, moderate, severe NPDR [19]. The methods used by Singh et al. utilized image analysis techniques for the early and automated detection of DR. This was completed with the assistance of other techniques such as image processing [20]. According to Zhao et al. [21] in their research, a novel saliency-based approach has been developed for the detection of fluorescence leakage in the angiography picture. According to their approach, which is supported by two publicly available datasets, diabetic retinopathy and malarial retinopathy, their findings are verified. Prasad et al. [22] advocated the use of morphological processes, such as word stemming, together with segmentation algorithms for finding blood vessels and microaneurysms. Principal component analysis is used for feature selection with greater focus on ease of use. In addition, back-propagation NN and the one-rule classifier approaches were utilized for the classification of photos to identify whether they were non-diabetic or diabetic. Akram et al. [23] built a hybrid classifier that works in two stages. The first step involves preprocessing images, and the subsequent steps entail the discovery of eye lesions as well as the development of various feature sets. The work is likely to lead to further expansion of m-Mediods modelling methodology, which consists of a Gaussian Mixture Model linked to Gaussian Mixture Model to result in a hybrid classifier designed to increase classification accuracy. Winder et al. [24] conduct a survey utilizing image recognition algorithms to aid in the automatic detection of retinopathy. Additionally, digital colour retinal images are used in the study. there were stages 1, 2, 3, 4, and 5, each representing different algorithm approaches (preprocessing, localization and segmentation of the optic disk, segmentation of the retinal vasculature, localization of the macula and fovea, localization and segmentation of retinopathy). The work "Automatic Extraction of Anatomical Features from Retinal Images for Early Glaucoma Detection" by Muhammad Salman et al. [25] reviews advanced approaches for the automated extraction of anatomical features in order to aid in the early diagnosis of glaucoma. These estimations evaluated the qualities which include CDR (optic cup to disc ratio), RNFL (retinal nerve fiber layer), and PPA (parapapillary atrophy) among others, as well as a critical assessment of currently available automatic extraction procedures. This is an additional value for the efficient feature extraction approaches in relation to the glaucoma diagnosis. Carrera et al. [26] developed a paradigm for early DR detection with digital retina image processing. The study's goal was to determine what sort of non-proliferative DR imaging diagnoses can be assigned. In order to extract features that might be used by SVM to categorize DR pictures, the pictures were first treated to eliminate blood vessels, hard exudates, and microaneurysms. The suggested model consisted of 400 captioned photos, each accompanied by an additional information box. According to the results, the proposed design has a sensitivity of 95% and a predictive proficiency of 94%. In the DRNP classification challenge, the generated model did well and had the potential to serve as a clinical prototype. A potential use of text analysis may be the use of it to improve the detection and identification of retinopathy.

He W et al. [27] introduced a previously unknown method for the quantitative study of multispectral images' spatial and heinous highlights of diabetic damage division. The proposed technique used the network approach's generalized low-rank estimation in conjunction with a strategy of directed regularization to examine ghastly highlights in all ghostly cuts. The Chetoui et al.'s proposed methodology may aid in the diagnosis of haemorrhages, exudates, and microaneurysms [28]. The team of researchers headed by Mookiah et al. [29] has compared several methods of DR imaging detection. Overall, a comprehensive analysis of numerous detection techniques for DR image diagnostics has been discovered from the research. The paper listed all of the techniques and the percentage of their accuracy. Eswari et al. [30] identified that the Bayesian optimization SVM is used to predict glaucoma in diabetic patients and is integrated into a local real-time diabetic population dataset to obtain an accuracy of 96.6%.

7.4 Methodology

7.4.1 Dataset

The dataset comprises 3662 shaded RGB anatomical design photography pictures in PNG design, freely available for the general public on Kaggle. All photographs are labelled with a five-point scale indicating the severity of DR: zero indicates no DR, one indicates mild, two indicates moderate, three indicates severe, and four indicates proliferative DR. The photos' mind-dominated portion lacks or has a moderate amount of DR. The dataset can be classified as per the class levels represented in Table 7.2.

Observations

- Although the test dataset is rather small, the train dataset is too.
- The dataset used for preparation is substantially larger than the dataset used for testing.

TABLE 7.2

Classification of Dataset as Per Image Representation

Class Label	Number of Images	Percentage of the Class in the Dataset
Class 0	1805	49.29%
Class 1	370	10.10%
Class 2	999	27.28%
Class 3	193	5.27%
Class 4	295	8.06 %

- Residual confounding in model building due to an excessive amount of fitting.
- The desire for transfer learning and further information grows.
- Clinicians have graded each image on a 0–4 scale according to the degree of DR. The issue is a multi-class problem with five distinct categories.
- That dataset is highly unbalanced.
- With the presence of no DR, there are multiple times as many images with this condition as images with severe conditions.
- The necessity of increasing the class load.
- DR is "0," "mild, or "moderate" severe, proliferative DR.

7.4.2 Instruction for the Grading

When grading the images, the graders were instructed to give all photos an assessment using the following rubric by *Grading Quality Control section*:

Phase 1: Quality of the image. Choosing the picture quality for each image will affect the overall image quality. The following image quality factors must be examined when reviewing each image:

- Grade: Is the grading adequate for minor retinal lesions, such as those with macular atrophy (MA, IRMA)?
- Illumination: If the image is too black, it can be because it lacks illumination. Are there any shading or clarity issues that have impact on the clarity of grading?
- Image field definition: Does the entire optic nerve head and macula belong to the primary field?
- Artifacts: Does the photo show enough of the object to offer an accurate grade?

Phase 2: Classification of the image.

- Excellent: No image quality factors concerns. All gradable injuries of retinopathy.
- Good: 1–2 quality factors problem. All gradable injuries of retinopathy.
- Sufficient: 3–4 image quality elements problems. All gradable injuries of retinopathy.
- Inadequate for detailed assessment (e.g., neovascularization indicated that there was so probable PDR but that the macula could be veiled so that DME cannot be rated).

- Not enough for any interpretation.
- Other: Some affected the rating of other performance metrics.
- No picture/technique issue.

Phase 3: Severity of DR/DME. Individuals are prompted to take the DR rating if the image is gradable:

- Choose the DR rating based on the associated international scale for clinical retinopathy.
- If hard exudate existed within one disc of the centre of the macula, select a referable macular diabetic (DME) oedema.
- Only select a grade DR/DME if the image is partially gradable.
- Note any noteworthy image results, even if it is not reproducible.

7.4.3 Development of the Algorithm

The technique of training the system to execute a specific task is called deep learning. It computes the severity of DR from the pixel intensities in a fundus picture. This function can only be "trained" with an extensive set of photos, and in the case of macular degeneration, the severity has already been known (training set). Neural networks have an initial setting where parameters (a mathematical function) are randomly chosen. To determine how severe the images are, the error on those images is compared to the known error in the training set, and the parameters of the function are slightly adjusted to reduce the discrepancy. The method is performed on every image in the training set numerous times, and after many iterations, the function "needs to learn" how to reliably determine the DR severity. The result is a function general enough to compute DR severity on new photos, provided that a patient's training data are available. This study used a CNN. Although the algorithm does not use lesions (e.g., haemorrhages, microaneurysms) as explicit indicators, it is quite likely that it will learn to distinguish them by leveraging the relevant local features [8].

7.4.4 Purpose of CNN

Additionally, there should be at least one convolutional layer, and at least one totally associated layer as in a normal multi-facet neural organization. CNN's engineering is centred on utilizing the 2-D building of an information picture to the fullest extent possible. It is possible to accomplish this by interpreting highlights that remain invariant [31]. CNNs, which are also known as recurrent neural networks, are neural networks that use a mathematical

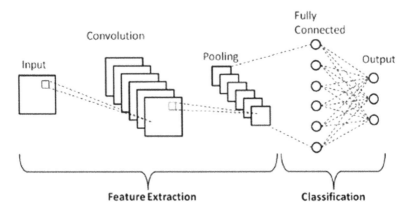

FIGURE 7.6
CNN architecture.

procedure known as convolution. CNNs, a type of neural network used to diagnose DR, have recently been used to analyze fundus images in order to establish their effectiveness at both detection and classification tasks [32]. While CNNs have the advantage of being simpler to train and needing fewer parameters than fully linked networks with the same number of hidden units, fully linked networks with the same number of hidden units may be more reliable. Figure 7.6 represents the architecture of the CNN. The layer convolutional network is good at predicting features relevant to an image in the previous layer, resulting in a hierarchy of non-linear features that gets more complex as the number of layers increases. All of these generated features are being used by the last layer(s) for categorization [33].

CNN deep learning improves model performance by progressively simulating small amounts of information, often sequentially over time, a process known as deep learning. A CNN is more accurate due to multilayer processing and its ability to learn to extract more significant characteristics from a series of images over time.

7.4.5 EfficientNet

The EfficientNet is a CNN architecture and scaling method that equally scales all dimensions of depth/width/resolution, which is achieved by the use of a compound coefficient. Unlike current practice, which uses arbitrary scales for these three elements, the EfficientNet scaling method consistently scales network breadth, depth, and resolution with a set of preset scaling coefficients. Consider an example where we would like to increase the number of computational resources by 2^N times. We may accomplish this by increasing

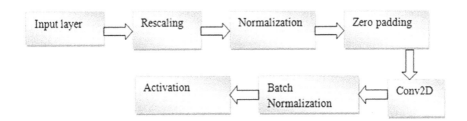

FIGURE 7.7
EfficientNet stem architecture.

the network depth by α, width by β, and image size by γ^N, where α, β, and γ are constants to be chosen using a tiny grid search on the initial little model. EfficientNet uses a compound coefficient φ to equally scale network width, depth, and resolution, which is an optimal approach to do so [34].

The compound scaling approach is supported by the assumption that if the input image is larger, then the network needs more layers to extend the receptive field and more channels to better differentiate fine-grained patterns on the larger image. In order to create a family of networks, the researchers first increased the size of the existing baseline network. Figure 7.7 illustrates the work flow of the EfficientNet.

- It all begins with the stem.
- In most cases, the models are either too large, are too deep, or have an extremely high resolution. Increasing these qualities initially improves the model, but it soon saturates, and the model created simply has more parameters and is thus inefficient. They are scaled in a more systematic manner in EfficientNet, i.e. everything is gradually raised.
- The structure is made up of seven blocks. These blocks also have a different number of sub-blocks, with the number of sub-blocks increasing with the number of sub-blocks.
- Finally, there are the last layers.

7.4.6 Evolution of Algorithm

The trained neural network generates a continuous number between 0 and 1 for referable DR and other DR classifications, corresponding to the probability of that condition being present in the image. Receiver operating curves were plotted by varying the operating threshold and two operating points for the algorithm were selected from the development set. The first operating point approximated the specificity of the ophthalmologists in the derivation set for detecting referable DR 96% and allowed for better comparison between the algorithm's performance and that of the five or seven ophthalmologists that graded the validation set. The second operating point corresponded to a sensitivity of 95% for detecting referable DR because a high sensitivity is a prerequisite in a potential screening tool.

7.4.7 Performance Analysis

With automatic retinal image analysis, the segmentation performance is often measured by checking whether or not each pixel in the segmented output image matches the ground truth image. A binary judgement is made; whether a pixel is classified correctly (true positive [TP] and true negative [TN]) or erroneously (false positive [FP] and false negative [FN]). If the retina is seen as a vessel, then this is known as the ROI, and this area is made up of the various parts of the retina, which include arteries, arterioles, capillaries, venules, or exudates. The classification error for the FP pixels is incorrect: They should be counted as belonging to the ROI. Non-foreground pixels (FN pixels) are those that are mistakenly assigned to the background. Other performance measures can be developed from these four metrics, which provide a comprehensive view of performance. Figure 7.8 illustrates the classification metrics. These include sensitivity (SE), true positive rate (TPR), or recall (Re), specificity (SP), false positive rate, positive predictive value (PPV), or precision, negative predictive value, and overall accuracy.

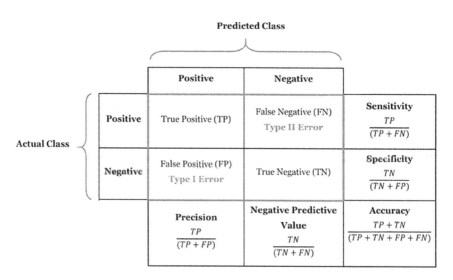

FIGURE 7.8
Confusion matrix with classified metrics.

7.5 Result and Discussion

7.5.1 Data Analysis

In this study, it was revealed that they could detect DR referred to as referable diabetic retinopathy (RDR) and vision-threatening retinopathy called VDR by employing a deep learning architecture which was equipped with a newly built CNN called EfficientNet. Two databases that were available in the tests, EyePACS and APTOS 2019, were used. In the process of automating the neural network creation, the technique for baseline networks that incorporates the neural architecture search is utilized. On a FLOPS (floating-point operations per second) basis, it is able to provide high accuracy and good efficiency. The movable inverted bottleneck convolution is implemented in this architecture (MBConv). The researchers then grew this baseline network in size in order to create a family of networks. Figure 7.9 represents the demographic analysis of images that can be classified into each class.

A significant advantage of our approach is that it makes us less vulnerable to attack. Technologies that have the potential to have a substantial impact on the disease's social and economic aspects allow general practitioners and other caregivers to promptly identify DR, which enables them to treat and care for people who are suffering from the condition, whether or not they have insurance. The earlier people can recognize DR, the better they will be able to keep their eyesight while also letting doctors concentrate on treatment alternatives. Figure 7.10 demonstrates that the data analysis of 67% of the complete dataset was utilized for training data, while 33% was utilized for testing data.

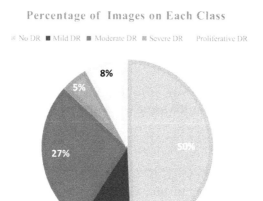

FIGURE 7.9
Spilt up the images on each class.

FIGURE 7.10
Data analysis.

7.5.2 Convolutional Neural Network

Object detection has been an area of machine learning study for many years. However, object separation is affected by parameters such as segmentation, lighting, and deformation. Due to the fact that CNNs are capable of recognizing a wide variety of objects, including handwritten digits, three-dimensional objects, and any other form of an object, the name they were given makes perfect sense. During the 2012 ILSVRC-2012 competition, CNN revealed its actual value by demonstrating their ability to assist in the image classification task using the ImageNet dataset, which contains 1.2 million high-resolution training images.

- In the first place, I built a gauge model to measure the previous score and enhance model-building fundamentals.
- Straightforward design.
- Using a fixed 2×2 bit size with ReLU activation and 2×2 convolutional layers provides the same effect as with ReLU initiation.
- Using a fixed part size of 2×2, two Max pooling layers are used.
- One dense layer per dense layer.

Data used in this algorithm may be found on Kaggle. In addition, the 81/81 Epoch value has a specific function. Gaussian filter could be used in tandem with the range 224×224. We can get 93.27% accuracy out of this.

7.5.3 EfficientNet Model

Developing a deep learning architecture based on a recently created CNN called EfficientNet that may be used to detect RDR and vision-threatening

DR. Two publicly available datasets were used for the tests: EyePACS and APTOS 2019. Here, too, we're utilizing the Kaggle dataset, which has an Epoch value of 92/92. Gaussian filter with the same range 224 × 224 can be used. However, the precision is 99.03%.

7.5.4 Model Prediction

The performance efficiency of the CNN and the EfficientNet model was evaluated using fundus images. The model achieved 93.27% and 99.03% accuracy, respectively, when fine-tuning the ConvNet setup on CNN and EfficientNet. On fundus images, the exhibition effectiveness of the convolutional neural organization and EfficientNet model was evaluated. The model achieved 93.27% and 99.03% accuracy on CNN and EfficientNet, respectively, in a calibrated ConvNet setup. As illustrated in Figure 7.11, the model preparation charts depict the preparation and approval exactness and misfortune values. In Figure 7.11a, the pattern of precision is ascending, whereas

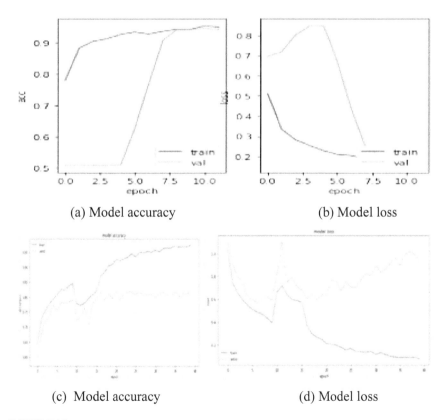

(a) Model accuracy   (b) Model loss

(c) Model accuracy   (d) Model loss

FIGURE 7.11
ConvNet configuration training graphs: (a) training accuracy graph CNN; (b) training loss graph on CNN; (c) training accuracy graph on EfficientNet; (d) training loss graph on EfficientNet.

in Figure 7.11b, the pattern of model misfortune or blunder is predominantly descending. Additionally, Figure 7.11c depicts a rising precision pattern for the efficient net, whereas Figure 7.11d depicts the preparation loss for the effective net, which is similar to Figure 7.11b. As we can see for module 0, the precision pattern expands as the number of ages increases. The greater the number of ages, the more learning occurs. Along these lines, we have a precision of 93.27% after ten ages and 99.03% after forty ages. Effective nets exhibit a similar pattern.

7.5.5 Accuracy

In order to build our model, we used preprocessing techniques from Kaggle on a fundus dataset provided from Kaggle and applied GoogleNet CNN architecture to the process. In Figure 7.12, the results are presented for 15,000 EyePACS fundus photos, which demonstrate a validation accuracy of 93.27% and 99.3% for a four-grade classification (no DR, mild/moderate NPDR, severe NPDR, and PDR). Table 7.3 demonstrates the proposed CNN and EfficientNet different validation performance results.

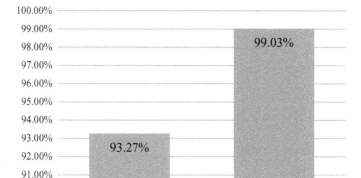

FIGURE 7.12
Accuracy performance.

TABLE 7.3

Performance Metrics

Algorithms	Accuracy (%)	Sensitivity (%)	Specificity (%)
CNN	93.27	97.6	82.59
EfficientNet	99.03	98.65	97.25

7.6 Conclusion

This groundbreaking research provided an AI-based cutting-edge deep learning algorithm for classifying DR from photos. To aid automated localization and detection of diabetes-related eye disorders in retinal images, such as glaucoma, DR, and DME, a new methodology is introduced in this paper that combines CNN EfficientNet and a training algorithm for solving image retrieval problems. The proposed methodology is made up of two phases: image categorization using severity as a baseline and model accuracy and loss prediction. In this research, we've presented a new method for diabetes detection that makes use of a highly effective network and achieves an accuracy of 99.03%. It is predicted that the model EfficientNet has a much better prediction than the convolutional neural network. The project's future goals include research in addition to these ailments, for example, cataracts, age-related macular oedema, and other eye ailments such as choroidal folds and cystoid macular oedema.

Acknowledgement

The authors would like to thank the management of Kalasalingam Academy of Research and Education for providing fellowship to carry out the research work.

References

[1] NCD Risk Factor Collaboration (2016) "Worldwide trends in diabetes since 1980: A pooled analysis of 751 population-based studies with 4.4·million participants". *The Lancet*, *387*(10027), 1513–1530

[2] Eswari, M. S., Karkuzhali, S. (2020, January) "Survey on segmentation and classification methods for diagnosis of glaucoma". In *2020 International Conference on Computer Communication and Informatics (ICCCI)* (pp. 1–6). IEEE.

[3] Murray, C. J., Lopez, A. D. (1997). "Mortality by cause for eight regions of the world: Global burden of disease study". *The Lancet*, *349*(9061), 1269–1276.

[4] Klein, R., Klein, B. E. (2013). "The prevalence of age-related eye diseases and visual impairment in aging: Current estimates". *Investigative Ophthalmology & Visual Science*, *54*(14), ORSF5-ORSF13.

[5] Wong, W. L., Su, X., Li, X., Cheung, C. M. G., Klein, R., Cheng, C. Y., Wong, T. Y. (2014). "Global prevalence of age-related macular degeneration and disease

burden projection for 2020 and 2040: A systematic review and meta-analysis". *The Lancet Global Health*, 2(2), e106–e116.

[6] Jager, R. D., Mieler, W. F., Miller, J. W. (2008). "Age-related macular degeneration". *New England Journal of Medicine*, 358(24), 2606–2617.

[7] Vo, H. H., Verma, A. (2016). "New deep neural nets for fine-grained diabetic retinopathy recognition on hybrid color space". *Int. Sym. on Multimedia. IEEE Conf.*, San Jose, CA, pp. 209–215.

[8] Ting, D. S. W., Cheung, G. C. M., Wong, T. Y. (2016). "Diabetic retinopathy: Global prevalence, major risk factors, screening practices and public health challenges: A review". *Clinical & Experimental Ophthalmology*, 44(4), 260–277.

[9] Dhoot, D. S., Baker, K., Saroj, N., Vitti, R., Berliner, A. J., Metzig, C., Singh, R. P. (2018). Baseline factors affecting changes in diabetic retinopathy severity scale score after intravitreal aflibercept or laser for diabetic macular edema: Post hoc analyses from VISTA and VIVID. *Ophthalmology*, 125(1), 51–56.

[10] Mohammadian, S., Karsaz, A., Roshan, Y. M. (2017). *Comparative study of fine-tuning of pre-trained convolutional neural networks for diabetic retinopathy screening. 24th National and 2nd Int. Iranian Conf. on Biomedical Engineering*, Tehran, pp. 1–6.

[11] Honnungar, S., Mehra, S., Joseph, S. (2016). *Diabetic retinopathy identification and severity classification*. Stanford University.

[12] Prasad, D. K., Vibha, L., Venugopal, K. R. (2015). "Early detection of diabetic retinopathy from digital retinal fundus images". *IEEE Conf. on Recent Advances in Intelligent Computational Systems*, Trivandrum, pp. 240–245.

[13] Asare, S. K., You, F., Nartey, O. T. (2020). "Efficient, ultra-facile breast cancer histopathological images classification approach utilizing deep learning optimizers". *International Journal of Computer Applications*, 11, 9.

[14] Lin, J., Yu, L., Weng, Q., Zheng, X. (2019). "Retinal image quality assessment for diabetic retinopathy screening: A survey". *Multimedia Tools and Applications*, 1–27.

[15] A Comprehensive guide explaining diabetic retinopathy and the treatments n.d. available: https://www.eye7.in/retina/diabetic-retinopathy/complete-guide/

[16] Diabetic retinopathy detection n.d.: https://medium.com/analytics-vidhya/diabetic-retinopathy-detection-2c6e0edcebb6

[17] Shankar, K., Elhoseny, M., Chelvi, E. D., Lakshmanaprabu, S. K., Wu, W. (2018). "An efficient optimal key based chaos function for medical image security". *IEEE Access*, 6, 77145–77154.

[18] Shankar, K., Lakshmanaprabu, S. K., Gupta, D., Maseleno, A., De Albuquerque, V. H. C. (2020). "Optimal feature-based multi-kernel SVM approach for thyroid disease classification". *The Journal of Supercomputing*, 76(2), 1128–1143.

[19] Raman, V., Then, P., Sumari, P. (2016, June). "Proposed retinal abnormality detection and classification approach: Computer aided detection for diabetic retinopathy by machine learning approaches". In *2016 8th IEEE International Conference on Communication Software and Networks (ICCSN)* (pp. 636–641). IEEE.

[20] Singh, N., Tripathi, R. C. (2010). "Automated early detection of diabetic retinopathy using image analysis techniques". *International Journal of Computer Applications*, 8(2), 18–23.

[21] Zhao, Y., Zheng, Y., Liu, Y., Yang, J., Zhao, Y., Chen, D., Wang, Y. (2016). "Intensity and compactness enabled saliency estimation for leakage detection in diabetic and malarial retinopathy". *IEEE Transactions on Medical Imaging*, 36(1), 51–63.

[22] Prasad, D. K., Vibha, L., Venugopal, K. R. (2015, December). "Early detection of diabetic retinopathy from digital retinal fundus images". In *2015 IEEE Recent Advances in Intelligent Computational Systems (RAICS)* (pp. 240–245). IEEE.

[23] Akram, M. U., Khalid, S., Tariq, A., Khan, S. A., Azam, F. (2014). "Detection and classification of retinal lesions for grading of diabetic retinopathy". *Computers in Biology and Medicine, 45,* 161–171.

[24] Winder, R. J., Morrow, P. J., McRitchie, I. N., Bailie, J. R., Hart, P. M. (2009). "Algorithms for digital image processing in diabetic retinopathy". *Computerized Medical Imaging and Graphics, 33*(8), 608–622.

[25] Haleem, M. S., Han, L., Van Hemert, J., Li, B. (2013). "Automatic extraction of retinal features from colour retinal images for glaucoma diagnosis: A review". *Computerized Medical Imaging and Graphics, 37*(7–8), 581–596.

[26] Carrera, E. V., Gonzalez, A., Carrera, R. (2017). "Automated detection of diabetic retinopathy using SVM". *24th Int. Conf. on Electronics, Electrical Engineering and Computing,* Cusco, Peru, pp. 1–4.

[27] He, Y., Jiao, W., Shi, Y., Zhao, B., Zou, W., et al. (2019). "Segmenting diabetic retinopathy lesions in multispectral images using low-dimensional spatial-spectral matrix representation". *IEEE Journal of Biomedical and Health Informatics, 24*(4), 1.

[28] Rajan, K., Sreejith, C. (2018). "Retinal image processing and classification using convolutional neural networks". *International Conference on Computational Vision and Bio-Engineering,* Palladam, vol. 30, Springer, pp. 1271–1280.

[29] Mookiah, M. R. K., Acharya, U. R., Chua, C. K., Lim, C. M., Ng, E. Y. K., Laude, A. (2013). "Computer-aided diagnosis of diabetic retinopathy: A review". *Computers in Biology and Medicine, 43*(12), 2136–2155.

[30] Eswari, M. S., Balamurali, S. (2021, March). "An intelligent machine learning support system for glaucoma prediction among diabetic patients". In *2021 International Conference on Advance Computing and Innovative Technologies in Engineering (ICACITE)* (pp. 447–449). IEEE.

[31] Sandler, M., Howard, A., Zhu, M., Zhmoginov, A., Chen, L. C. (2018). "Mobilenetv 2: Inverted residuals and linear bottlenecks". In *Proceedings of the IEEE Conference on Computer Vision and Pattern Recognition* (pp. 4510–4520).

[32] Lim, G., Lee, M. L., Hsu, W., Wong, T. Y. (2014, June). "Transformed representations for convolutional neural networks in diabetic retinopathy screening". In *Workshops at the Twenty-Eighth AAAI Conference on Artificial Intelligence.*

[33] Kulkarni, P., Zepeda, J., Jurie, F., Perez, P., Chevallier, L. (2015, April). "Hybrid multi-layer deep CNN/aggregator feature for image classification". In *2015 IEEE International Conference on Acoustics, Speech and Signal Processing (ICASSP)* (pp. 1379–1383). IEEE.

[34] Tan et al. n.d. *Efficient Net: Rethinking model scaling for convolutional neural networks:* https://paperswithcode.com/

8

Protecting and Analyzing Big Data on Cloud Platforms

Avita Katal, Niharika Singh, Vitesh Sethi, and Susheela Dahiya

University of Petroleum and Energy Studies, Dehradun, India

CONTENTS

8.1 Introduction to Big Data and Cloud Computing 116
    8.1.1 Cloud Computing Service Models 117
    8.1.2 Deployment Models of Cloud 118
8.2 Big Data and Cloud Relationship 119
    8.2.1 Hadoop: The Big Data Software 119
    8.2.2 Models Between the Big Data and Cloud 120
8.3 Need for Retrieving Information from Widespread Big Data
    on the Web 120
8.4 Security of Big Data in Cloud Computing 121
8.5 Challenges for Information Retrieval in Big Data and Cloud 122
8.6 Classification of Big Data Security 123
    8.6.1 Infrastructure Security 123
    8.6.2 Data Privacy 124
    8.6.3 Data Management 125
    8.6.4 Integrity and Reactive Security 125
8.7 Encryption Decryption Algorithm for the Analytics
    of Big Data in Cloud 125
    8.7.1 Homomorphic Encryption Algorithm 125
    8.7.2 Verifiable Computation Algorithm
        (Outsource Computing) 126
    8.7.3 Message Digest Algorithm 126
    8.7.4 Key Rotation Algorithm 127
    8.7.5 Data Encryption Algorithm (DES) Algorithm 127
    8.7.6 Rijndael Encryption Algorithm 127
8.8 Advanced Security by Homomorphic Cryptosystems 128
    8.8.1 Fully Homomorphic Encryption 129
    8.8.2 Partially Homomorphic Encryption 129

DOI: 10.1201/9781003134138-8

8.9   Case Study with Results 131
    8.9.1   AES Encryption Performance 132
         8.9.1.1   Algorithm 132
         8.9.1.2   Speed 132
    8.9.2   ElGamal Encryption Performance 133
         8.9.2.1   Algorithm 133
         8.9.2.2   Speed 133
8.10 Conclusion 137
References 138

8.1 Introduction to Big Data and Cloud Computing

The word cloud in cloud computing is a metaphor for "Internet"; therefore, cloud computing is a kind of Internet-based computing in which services like storage, computers, and servers are given through the Internet [1]. Cloud computing is the powerful technology that is able to do large computations. It has many advantages like data service integration, parallel processing, scalable data storage and security, and virtualization. It not only minimizes the expenses for computerization by users and enterprises but also minimizes the expense of maintaining the infrastructure. Virtualization is the most important technology that implements cloud computing.

Big Data refers to the large volume of data that is difficult to process, store, and analyze through old database techniques. Big Data is called as the datasets with very large volume such that the dedicated software that is used today is not able to manage and process it in a particular time period. Data is not restricted to the structured data but also includes the unstructured data. Big Data needs a huge storage space. A typical storage of Big Data infrastructure is based on the cluster network attached storage (NAS). The infrastructure of clustered NAS needs properties of different NAS "pods" with each NAS "pod" consisting of the different devices for the storage linked to the NAS device. The NAS devices are linked with each other to enable large searching and sharing of data [2].

The storage of data through the cloud is a better choice for the small-scale enterprises that consider the utilization of Big Data analytic techniques. The owners of the small-scale business are not able to afford the clustered NAS technology so they can look for the number of models of cloud computing to overcome the needs of Big Data.

The 5v's of Big Data are shown in Figure 8.1 and are described below [3]:

- *Volume* – It is the amount of data collected through various sources. The advantage of collecting huge data includes the generation of valuable information and patterns by analysis of data.

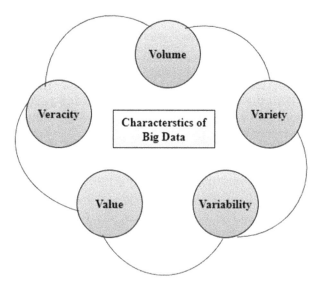

FIGURE 8.1
5V's of Big Data.

- *Variety* – It is known as the data generated by smartphones, sensors, or social media. This data includes text, data logs, video, image, audio, and any unstructured or structured type.
- *Velocity* – It is known as the data transfer speed. The data is changing continuously because of last archived data and absorption of complementary data collections.
- *Value* – It is known as the process of getting important data from huge datasets with rapid generation and different types.
- *Variability* – With rising variety and velocity of data, the flow of data is constant with periodic peaks. Event triggered peak, daily and seasonal loads of data are difficult to manage, even more when the data that is unstructured.

8.1.1 Cloud Computing Service Models

- Software as a Service (SaaS) – The cloud service provider (CSP) is responsible for providing various software applications to the clients and utilizing them without getting installed. SaaS is responsible for giving software for Big Data analysis. SaaS gives services to help the Big Data users to do data computations [4].
- Platform as a Service (PaaS) – CSP gives the tools, platforms, and other services to the users. The CSP is responsible for managing the middleware and the OS, with the services that allow it to provide

simple to sophisticated cloud-based applications. The benefits of PaaS are as follows: reducing risks by utilizing the pretested technologies, promoting the resources that are shared, better security for software, and minimizing different requirements that are required for the system development. In association with Big Data, PaaS gives a platform to the companies for making and utilizing customized applications that are required to analyze the huge amount of data that is unstructured at a less cost.

- Infrastructure as a Service (IaaS) – The CSP is responsible for providing the infrastructure like storage, computing capacity, etc. In the IaaS model, any outsourced organization delivers storage, hosts servers, hardware, and the infrastructure of devices to the users. The benefits of IaaS include business agility, increased security, and increased financial flexibility [5, 6].

- Data as a Service (DaaS) – It differs from old models like IaaS, SaaS, and PaaS in terms of providing data to the clients by the network [7] in conjunction with cloud on the basis of solving the problems in the management of data. For the above-mentioned reasons, DaaS is associated with Big Data whose services should be utilized [8]. DaaS is responsible for providing data processing and distribution methods. DaaS is similar to SaaS and storage as a service that can be merged with both the models [9].

8.1.2 Deployment Models of Cloud

There are three kinds of clouds: the private cloud, the public cloud, and hybrid cloud. A public cloud can be used by anyone [10]. In the public cloud, the enterprises do not own the core technology services and resources as the entire resources are managed by external organizations [11]. A public cloud can also be considered as the external cloud [12].

A private cloud can be known as the internal cloud for the business and not available to the general public for use. In the private cloud, the enterprises own entire services and resources. Services are available in enterprise through intranet. Since it is managed and owned by the enterprise, private cloud is costlier than the public cloud.

The hybrid cloud is a mixture of private and public cloud. A hybrid cloud enables the enterprises to store important data inside the firewall but using the public cloud for unimportant data.

Private cloud provides the most effective model for the analysis of Big Data, while preparing the internal resources with the services of public cloud. The hybrid cloud provides the enterprises to utilize the on-demand storage services for particular analysis initiatives via services of public cloud by providing the scale and capacity as required by the customer.

8.2 Big Data and Cloud Relationship

With the advancement of technology, rapid development of electronic information society has been seen which also has led to increase in utilization of cloud, problem in dealing with Big Data, etc. [13]. Big Data and cloud computing are rising side by side. Cloud computing is evolving day by day giving solutions for the Big Data [14] as the storage services that are traditional were not able to tackle the problems of Big Data. Cloud computing overcomes the problems like huge amount of data and heterogeneity by enabling distribution of data, i.e., storing of data in more than one availability zone. Clouds are generally built for general purpose workloads, and pooling of resources in cloud environments provides much flexibility needed for Big Data.

Big Data storage and processing requires the expansion in the resources and this can be done through cloud computing as it provides the expansion of resources through virtual machines and helps in evolving and accessing the Big Data. Amazon, Google, and IBM are some examples that use Big Data in the cloud environment [15]. The cloud computing system should be modified so that they can easily fit with the Big Data requirements [16]. A resource pool with special computer chips for high-performance computing for Big Data should be made or the computational power can be increased by having CPUs that have multi cores. Also, special considerations are needed for ingesting the data into a Hadoop cluster, with a dedicated network that enables parallel processing algorithms, like MapReduce, to distribute the data among the nodes. Use of a special type of storage system that enables accessing and storing data as objects instead of files are some of the changes that can be brought to the cloud environment to suit Big Data.

8.2.1 Hadoop: The Big Data Software

Big Data processing has seen an increased interest with the availability of new software known as Apache Hadoop. It is a java-based distributed system, developed by Apache that enables the distributed computations of large data sets among clusters of computers by utilizing models of programming. It is developed in such a way that it is scalable from a single server to the thousand servers in which every server provides the local storage and computation. The data is usually generated from various services and formats. It has a distributed file system known as Hadoop Distributed File System (HDFS) that is capable of storing the information in various servers with different functions. The main property of Hadoop is an open-source implementation of MapReduce.

It is a programming model that allows computations of large datasets. It provides system for computation of large datasets by the use of following functions:

- The map function that computes the key/value pair required to make a set of intermediate key/value pairs.
- The MapReduce function, that computes the intermediate values generated and combines them to give a solution.

8.2.2 Models Between the Big Data and Cloud

The most commonly used models for the analytics of Big Data on cloud are SaaS and PaaS. IaaS cannot be utilized with the high-level applications of data analytics [20] but can be used to tackle the computation requirements and storage necessities of data. Cloud computing is a kind of distributed computing which gives services to user in the form of SaaS, IaaS, and PaaS, but with the rising use of Big Data, the model of cloud is shifting to the Big Database service (BDaaS) that includes Availability as a Service (AaaS). BDaaS called database as a service means the availability of services associated with database applications deployed in any implementation environment [17]. BDaaS is the same as SaaS or IaaS and it relies on the cloud storage for the constant access to the data. DBaaS is not in trending but refers to the outsourcing services and functions related to handling a large amount of data in cloud-based Big Data analytics [18]. Analytics as a Service (AaaS) gives the on-demand solutions to the clients and Model as a Service (MaaS) in which models are given as analytics solutions. MaaS and AaaS constitute services for the analysis of data analysis. However, these analytics services still face problems since it is difficult to calculate the reliability and quality of results of input data.

8.3 Need for Retrieving Information from Widespread Big Data on the Web

Million terabytes of data is getting stored on the web on a daily basis. With the advancement of cloud, most of the enterprises are storing their web data on cloud. The cloud-based Big Data analytics also helps in reducing the cost of data storage. This data can be used for retrieving useful information and can also play a major role in faster and better decision making. But the data present on the web cloud is meta data, which needs to be cleaned and filtered before use. This large amount of Big Data can be analyzed instantly to uncover hidden patterns, correlations, and other insights by using computational techniques. This Big Data analytics can also help the enterprises to identify a pattern in their customers' need which in turn help them in identifying new opportunities and can also help them in deciding the launch of new product.

There are many key technologies which can be used for Big Data analytics for retrieving useful information and discovering hidden patterns. Some of them are predictive analytics, machine learning, in-memory data fabric, Hadoop, data mining, text mining, etc. These technologies can be used separately or in combination to obtain the desired information for making strategic decisions for the growth of enterprises.

8.4 Security of Big Data in Cloud Computing

Cloud and Big Data are crucial stages for the development of the IT industry. Cloud is an open environment due to which the issues related to privacy and security are of major concern. The computation of data and storage of Big Data are adversely affected by the privacy and the security issues because of the usage of services provided by the third party which brings in new challenges.

The storage of data is done in a central location called cloud storage server where the data processing takes place, so that the client has trust on CSP and data security. The Service Level Agreement (SLA) should be standardized so that trust can be developed among the customer and the service provider. The cloud client data security varies in the requirement for the protection. Customers want protection of their data through login access controls while the intellectual property, classified, or structured data need the advanced security controls that include login, data hiding, encryption, etc.

The SLA reflects the contact between the service provider and the user on the quality of the services. It can be considered as a way to improve the level of security where various security levels and complexities can be identified depending on the services, to understand the rules of security of cloud customers, and to safeguard the data. Some rules are associated with SLA to safeguard the data, security, scalability, privacy, capacity, and the issues related to availability like data growth and data storage. The techniques used to secure Big Data, such as trap detection, encryption, and registry entry are very important. The Big Data security in cloud is necessary because of the problems given below:

- The Big Data security from threats and malicious intruders.
- The details about how CSP securely maintain space of the storage and delete the already present Big Data.

8.5 Challenges for Information Retrieval in Big Data and Cloud

When handling large amounts of data stored on cloud, the problems such as variety, verification, volume, and velocity arise. The different types of data involved in the analysis of Big Data are as follows: unstructured data, data not defined in a particular data model or not organized in a predefined manner; structured data that has predefined data format and length; mixed data, data from different sources and semi structured data, which does not follow the formal structure of data models of RDBMS. Figure 8.2 depicts the different varieties of data. In the domain of networking, the expenses of communications is a major problem as compared to the cost of data computations. The bandwidth and latency affect the computation of Big Data [19]. The major challenges faced during information retrieval from Big Data stored in cloud environment and their solution are as follows:

- Heterogeneity – The Big Data stored on cloud collected from different sources which makes it more heterogeneous in nature. The algorithms that are used for analysis expect homogenous data. As a first step in analysis of data, the data should be structured.

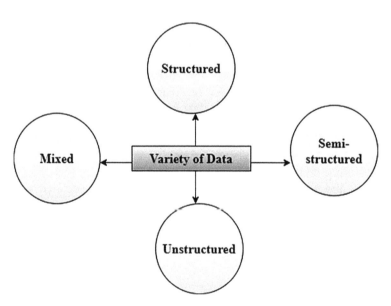

FIGURE 8.2
Variety of data.

- Data storage – The storage of large amounts of data by traditional techniques causes problems. The techniques used for the protection of data are not good and the velocity of Big Data needs the system for storage to increase rapidly, and that is very challenging through traditional systems of storage. The CSP provides space that can tolerate the faults and provide solutions that address the main challenges of storage of Big Data.

- Scale – It is the most crucial thing which means the size of Big Data. The management of the very huge data is the main problem for many decades. In the past, the challenge of large amounts of data was tackled by faster processors that followed Moore's law that provides us the resources needed to tackle the problem of large volumes of data. But there is a fundamental transfer underway, and the volume of data is increasing rapidly but the resources of the computer and the speed of the CPU are static.

- Data transfer – The data passes through the different steps like collection of data, input, processing, and output. Therefore, the data compression method should be employed to minimize the volume because it reduces the transfer speed. The concept of bringing the analytics to the data rather than bringing the data to the analytics can be a solution to the data transfer problem.

8.6 Classification of Big Data Security

The challenges are classified into four aspects of the Big Data ecosystem. Figure 8.3 shows the classification of Big Data security. These are [20]:

8.6.1 Infrastructure Security

It is important to mention the primary advancements and structures in securing the design of a Big Data system, especially those that are based on

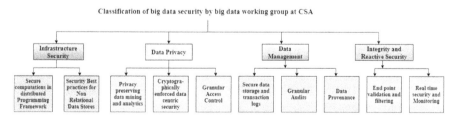

FIGURE 8.3
Classification of Big Data security.

Hadoop innovation. Hadoop is the accepted technology for actualizing the Big Data in an organization.

For instance, a G-Hadoop security model (an augmentation of the MapReduce structure to keep running on different groups) that simplifies some security components and customers verification for a particular aim to shield the framework from conventional attacks.

Another approach is that of depicting another architecture of big data, or changing the current architectures, by considering the end solution to enhance the security of the environment. The authors in [21] have proposed another architecture on the basis of Hadoop file system that can be joined with multi-node reading and network coding, making it conceivable to enhance the system security. The other arrangement concentrates on securing the group communications in extensive scale systems managed by systems of big data and can be accomplished by making certain conventions and modifying the node infrastructure. By focusing on authentication, architecture security, communication security, and availability, the infrastructure security of the Hadoop system can be improved. The major issues of infrastructure security can be overcome by changing the big data scheme and by the addition of new layers of security.

8.6.2 Data Privacy

Nowadays most people are concerned about data privacy. It is a crucial issue in the big data systems. A big data system has a huge amount of personal data that is used by organizations for providing customized personal services to each customer. Although organizations don't have the power to utilize the information of customer's without prior knowledge, they tend to have some advantages by utilization of customer data. Different types of techniques have been developed which safeguard the information and enable enterprises to use it. Such technologies are as follows:

- Cryptography – The massive amount of data is transferred from one place to another that may contain confidential information so encryption algorithms should be applied to protect the data from potential threats during transfer.

- Access control – It is a security technique that allows who or what can view the resources or information in a computing environment. In the case of big data, HDFS is utilized on private clusters behind the firewalls and it needs the strong authorization and authentication to safeguard the sensitive public and private data.

- Confidentiality – It refers to protecting information from unauthorized persons.

- Anonymization – It is technique for eliminating the personal information from the large datasets so that the identity of the person whom the data describes remains anonymous.

8.6.3 Data Management

Once the data is stored, another focus is on improving the different policies to improve sharing of data. As Big Data stores huge amounts of data, it is not only important to store data but also how to collect this data. The authors [22] have presented a solution to safeguard the privacy of the customer by making a specific parameter to calculate the level of privacy. Another technique used is distributing the data stored in the Big Data system into sequenced parts and storing them with various CSPs, thus making complete data unavailable at single source. Other implementations for data management include:

- Sharing algorithms – To get the maximum possible value from the data, it is important to distribute the data between the clusters.
- Laws, government, and policies – The government should make strict laws and policies for the security of data in the Big Data systems.

8.6.4 Integrity and Reactive Security

One of the main properties of Big Data is the capability to receive the data from different sources with different formats either in the structural or non-structural form which increases the necessity for checking the data integrity. Integrity is also known as maintaining the accuracy and consistency in the data. It safeguards the data from unauthorized modifications during its entire lifecycle.

- Attack detection – It is one of the most important security parameters in the context of Big Data because when attack is performed on the large amount of data, it is difficult to detect where the attack is happening.
- Recovery – The data recovery is one of the most important things in the context of Big Data. There should be proper recovery tools in case of any attack or any disaster in the cloud data centre.

8.7 Encryption Decryption Algorithm for the Analytics of Big Data in Cloud

8.7.1 Homomorphic Encryption Algorithm

It is an encryption algorithm that enables only some particular computations on plain texts and generates results in an encrypted form. Rivest–Shamir–Adleman (RSA) is the first algorithm that has a homomorphic property. In terms of algebra, it can be defined as the structure preserving map among the two algebraic structures such as groups.

8.7.2 Verifiable Computation Algorithm (Outsource Computing)

It allows a frail client to transfer data on the cloud without considering the issues related to security. This is one of the most secure algorithms that allows customers to transfer data to the cloud.

　Algorithm:

- VC = (KeyGen, Probgen, verify, compute) constitutes four algorithms as follows:
- KeyGen (L, lambda) -> (PK, SK): It provides two keys: private key and the public key on the basis of the parameter of the security that is lambda. The public key encrypts the l that is the target function and transfers it to the server for the private key that is held by the customer.
- ProbGenSK (x) -> (σa, τa): The algorithm that generates problems encrypts the input of function a into two values, private and public utilizing the secret key. The public value σa is provided to the provider to compute F (a), but secret value τa is handled by the client.
- ComputePK (σa) -> σb: The provider processes and encrypts value σb of the function's output b = F (a) utilizing the customer's PK and the input σa that is encoded.
- VerifySK (τa, σb) → b ∪⊥: The algorithm for verification converts the output σb that is encrypted by the client in the output of the function F utilizing the both SK and the secret "decoding" τa. It outputs b = F (a), if the σb represents an output that is valid of F on a, or outputs ⊥ otherwise.

8.7.3 Message Digest Algorithm

It takes an input of any length and gives a message digest, i.e., 128 bits long.

　It distributes the input in 512 bits of each block. The last block is inserted with 64 bits. The 64 bits are used to store the value of the original output. Extra bits are attached to the end, if the last block is less than 512 bits. Next, each block is distributed into 16 words of 32 bits each. MD5 utilizes a buffer that consists of four words each 32 bits long. MD5 utilizes table N that has 64 elements. Element number w is indicated as Nw. The table is processed before to increase the speed of computations. The elements are processed by utilizing the sin function as shown in Equation 8.1.

$$Nw = abs\big(sin\big(w+1\big)\big) * 232 \tag{8.1}$$

MD5 utilizes the four auxiliary functions. Each auxiliary function takes three 32-bit words as an input and gives the output as a 32-bit word.

After that, it applies the logical operators or, and, not, and xor to the input bits.

$$G(D,E,F) = (D \text{ and } E) \text{ or } (\text{not } (D) \text{ and } F) \tag{8.2}$$

$$H(D,E,F) = (D \text{ and } F) \text{ or } (E \text{ and not } (F)) \tag{8.3}$$

$$I(D,E,F) = D \text{ xor } E \text{ xor } F \tag{8.4}$$

$$J(D,E,F) = E \text{ xor } (D \text{ or not } (F)) \tag{8.5}$$

The contents of the four buffers (P, Q, R, and S) are then combined with the words of the input, utilizing the four auxiliary functions (G, H, I, J). These are the four rounds in which each round involves the 16 basic operations. The buffer P, Q, R, S has the MD5 digest of original input after the completion of all rounds. The four rounds are shown in Equation 8.2, 8.3, 8.4, and 8.5.

8.7.4 Key Rotation Algorithm

The storage and transference of data along the cloud environment can be done by using a shared secret key. The owner of the data allows to encrypt the data by utilizing the shared symmetric key. The data owner provides support to the cloud service provider that can convert the plain text to the cipher text and store in the cloud. The same steps should be followed in order to get the data from the cloud, decrypt, and be made available to the customer.

8.7.5 Data Encryption Algorithm (DES) Algorithm

It has a 64-bit size of the block during the running. It is specifically a 16-round FeistelCipher. During the communication, both receiver and sender must know the same secret key that can be utilized to decrypt and encrypt the data. The DES can be utilized for storage of data in an encrypted form.

8.7.6 Rijndael Encryption Algorithm

It is a block cipher algorithm that supersedes the DES. It is an algorithm for symmetric key encryption utilized to encrypt the data that is sensitive. It is designed on the basis of the three benchmarks:

- Protection from attacks;
- Compactness and speed of code on a wide range of platforms;
- Simple design.

8.8　Advanced Security by Homomorphic Cryptosystems

Homomorphic cryptosystems are basically used for data storage and security purpose. The encryption function used in such systems shows homomorphic nature and preserves operation which was performed by groups on ciphertexts. According to the property of cryptosystem, a third party is allowed to take two ciphertexts and perform task with it. The ability to perform basic calculations on ciphertexts allows different varieties of basics or simple security conventions to be based upon homomorphic cryptosystems. But due to the additional structure of the cryptosystem it limits the security.

Cryptosystems provide a mechanism which allows data integrity. In general case, cryptosystem gives a technique to change a message which is known as plaintext into another message (Ciphertext) with the help of some secret keys. If it is secure, then it is made public. According to the definition of cryptosystem, it consists of five different tuples (L, M, N, O, P) which follow certain conditions:

a. L represents finite set of plaintexts that are possible;

b. M represents finite set of ciphertexts;

c. N represents finite set of possible keys;

d. For every $N \in N$, an encryption lead on $\in O$ and a decrypting guideline $pn \in P$. Each on : $L \rightarrow M$ and $pn : M \rightarrow L$ are capacities with the end result that $pn\ (on\ (a)) = a$ for each plaintext $a \in L$

The main objective of cryptography is to provide security that can guarantee the secrecy of the framework. These objectives can be recorded under the following five classes:

- Authentication: The identity of the receiver and sender should be verified before sending the message.

- Confidentiality: Only authenticated or verified users can interpret the message, so that nobody else can utilize it.

- Integrity: The content of communication data is free from any type of modification.

- Service reliability and availability: Since gate-crashes influence their accessibility furthermore, clients. Framework ought to give an approach to give their clients the nature of administration they expect.

- Non-repudiation: This function implies that neither the sender nor the beneficiary can deny that they have sent a specific message.

There are different types of cryptosystem algorithms that can be used for different security purposes. These are as follows:

- Secret key cryptography: It employs a one key for decryption and encryption. To encrypt the plain text, the sender utilizes the key and transfers the encrypted text to the receiver. He utilizes the same key to get the decrypted data in the form of plain text.
- Public key cryptography: It utilizes different keys for decryption and encryption. It can be used for authentication, non-repudiation, and exchange of keys.
- Hash functions: It is also known as the message digest that does not use any keys. Instead of keys, a hash value of particular length is processed on the basis of plaintext that is highly secured and not possible to recover the contents or length of the plain text.

Table 8.1 shows the characteristics of existing homomorphic encryption cryptosystems.

8.8.1 Fully Homomorphic Encryption

The fully homomorphic ideas and notations were developed by Rivest, Adle-man, and Dertouzos after the invention of RSA. They requested an encryption work that licenses encoded information to be worked on without preparatory decoding of the operands, and they called those plans security homomorphism. It is a technique that supports encryption for both additive and multiplicative subsequently maintaining the ring structure of the plaintext space. Utilizing a scheme like that makes it possible to let an untrusted party do the calculations while never decoding the information, and accordingly safeguarding their protection. Figure 8.4 shows the homomorphic encryption scheme.

8.8.2 Partially Homomorphic Encryption

The idea of doing the simple and logical computations on the messages that are encrypted was first presented by Rivest, Adle-man, and Dertouzos. The first inspiration for this homomorphism was to take into account an encoded database to be put away by an outsider and to permit the proprietors and other approved individuals to perform counts with the information without decoding it. A multiplicative homomorphic cryptosystem has an encryption work A that fulfils the property as in equation 8.6:

$$A(B1) * A(B2) = A(B1 * B2) \tag{8.6}$$

Where B1 and B2 are plaintext.

TABLE 8.1

Characteristics of Existing Homomorphic Encryption Cryptosystems

Characteristics	RSA	Paillier	El Gamal	GoldwasserMicali	Boneh-GohNissim	Gentry
Platform	Cloud	Cloud	Cloud	Cloud	Cloud	Cloud
Type of homomorphic encryption	Multiplicative	Additive	Multiplicative	Additive, but it can encrypt only a single bit.	Only one multiplication, but unlimited number of additions.	Fully
Data privacy	Is guaranteed in communication and storage processes.	Is guaranteed in communication and storage processes.	Is guaranteed in communication and storage processes.	Is guaranteed in communication and storage processes.	Is guaranteed in communication and storage processes.	Is guaranteed in communication and storage processes.
Application of security	Cloud server	Cloud server	Cloud server	Cloud server	Cloud server	Cloud server
Keys used by	Customer	Customer	Customer	Customer	Customer	Customer

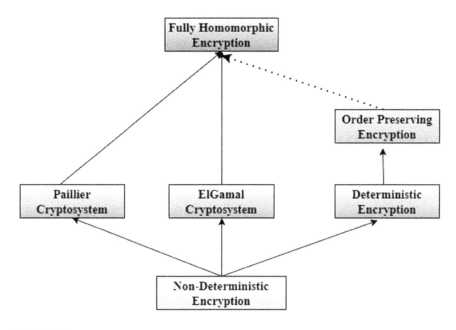

FIGURE 8.4
Homomorphic encryption scheme.

8.9 Case Study with Results

In this chapter, the implementation of security on Big Data over cloud is presented. With the vision to secure health data without making it difficult for the usual customers to share data through network. The encryption is applied on the data that is uploaded by the user in order to make the data more secure from brute force attack even by the cloud service provider. The encryption key itself was further encrypted with another algorithm. For making this final encrypted key completely unavailable to CSP, it was then partitioned in two random parts, one part shared socially and other made available only to the doctor's local machine.

Symmetric encryption is fast, is reliable, is easy to use, and uses less computer resources. First, the data was encrypted with private key symmetric encryption. And then this private key was further encrypted with ElGamal encryption.

8.9.1 AES Encryption Performance

8.9.1.1 Algorithm

AES process on bytes instead of bits. Therefore, the 128 bits of plain text are considered as 16 bytes by AES. These 16 bytes are put in four rows and four columns for processing as a matrix.

Following are the encryption steps of 128-bit block:

- get the round key set from the cipher key;
- give values along with block data to state array;
- add the first-round key to the starting state array;
- do nine rounds of state operation;
- do the tenth and the final round of state operation;
- print the final state array in cipher text.

8.9.1.2 Speed

The performance of the AES algorithm varies with the processor. It takes a huge amount of time even with the best hardware configurations to solve the advanced mathematics to break the code. Figure 8.5 shows the speed of AES.

AES Performance per CPU core for TLS v1.2 Ciphers
(Higher is Better, Speeds in Megabytes per Second)

	ChaCha20	AES-128-GCM	AES-256-GCM	AES-128-CBC	AES-256-CBC	Total Score
AMD Ryzen 7 1800X	573	3006	2642	1513	1101	≈ 8835
Intel i7-6700	585	2607	2251	1561	1131	≈ 8135
Intel i5-6500	410	1729	1520	1078	783	≈ 5520
Intel i7-4750HQ	369	1556	1353	688	499	≈ 4465
AMD FX 8350	367	1453	1278	716	514	≈ 4328
AMD FX 8150	347	1441	1273	716	515	≈ 4292
Intel E5-2650 v4	404	1479	1286	652	468	≈ 4289
Intel i7-2700K	382	1353	1212	763	552	≈ 4262
Intel i7-3840QM	373	1279	1143	725	520	≈ 4040
Intel i5-2500K	358	1274	1140	728	522	≈ 4022
AMD FX 6100	326	1344	1186	671	481	≈ 4008
AMD A10-7850K	321	1303	1176	685	499	≈ 3984
AMD A8-7600 Kaveri	306	1246	1108	648	470	≈ 3778
Intel E5-2640 v3	303	1286	1126	585	419	≈ 3719
AMD Opteron 6380	293	1203	1063	589	423	≈ 3571
AMD Opteron 6378	282	1138	986	561	406	≈ 3373
AMD Opteron 6274	232	1054	926	524	376	≈ 3112
Intel Xeon E5-2630	247	962	864	541	394	≈ 3008
Intel Xeon E5645	262	817	717	727	524	≈ 3047
Intel Xeon L5630	225	701	610	626	450	≈ 2612
Intel i7-2635QM	151	989	881	564	404	≈ 2989
AMD Opteron 2382	249	651	485	215	150	≈ 1750
Intel i7-950	401	256	218	358	257	≈ 1490
AMD Phenom 965	404	84	63	282	198	≈ 1031
Intel Core2 Q9300	231	126	133	221	161	≈ 872

FIGURE 8.5
Speed of AES.

8.9.2 ElGamal Encryption Performance

8.9.2.1 *Algorithm*

A change of ElGamal can be used to encrypt messages. The first step is to select a random *a* to encrypt a message, such that *a* is proportionately prime to *b* - 1. Then compute

$$s = ca \bmod b \tag{8.7}$$

$$t = daV \bmod b \tag{8.8}$$

The pair (s, t) is the ciphertext. The ciphertext is double the size of the plaintext. To decrypt s and t, compute

$$V = t / sz \bmod b \tag{8.9}$$

Since, $sz \equiv caz \pmod{b}$, and $t/sz \equiv daV/sz \equiv czaV/cza \equiv V \pmod{b}$, all this can be used. Equations 8.7 and 8.8 show the encryption equation and Equation 8.9 shows the decryption equation.

8.9.2.2 *Speed*

The speed of the ElGamal encryption depends upon the hardware of the system.

Figure 8.6 shows the ElGamal encryption.

	Public Key:
P	prime (can be shared among a group of users)
G	$<p$ (can be shared among a group of users)
Y	$= g^x \bmod p$
	Private Key:
X	$<p$
	Encrypting:
K	Chosen at random, proportionately prime to *p* - 1.
A	(ciphertext) $= g^k \bmod p$
B	(ciphertext) $= y^k M \bmod p$
	Decryption:

$$M \text{ (PLAINTEXT)} = B/A^X \bmod P$$

FIGURE 8.6
ElGamal encryption.

Figure 8.7 shows the key generated and performing encryption and Figure 8.8 shows the procedure for decryption.

```
Encrypting
Enter Symmetric Key of 8-10 digit long
9999009999
Private Symmetric Encryption Key : 9999009999
Encrypting file: Sample_report.pdf
Files encrypted successfully
Encrypting file: Test_File
Files encrypted successfully
Encrypted Private key: 35990263401006970509
Parts of patients private key :
 Social network part :359902634
 Doctor's part : 0100697059
```

FIGURE 8.7
Key generated and performing encryption.

```
For Decryption
Enter patient's key
359902634
Enter doctor's key
0100697059
Decrypting file: Sample_report.pdf
Decrypting file: Test_File
Files decrypted successfully
```

FIGURE 8.8
Performing decryption.

Social part of the key will be stored with data by CSP and shared online by the patient, whereas doctor's part of key will only be shared by the patient to the particular doctor with whom he wants to share his data. Figure 8.9 shows the encrypted data with key.

FIGURE 8.9
Encrypted data with key.

Figure 8.10 shows the ElGamal code.

```java
class ElGamal
{
    Random sc ;
    private String key;
    private BigInteger p, b, c, secretKey, brmodp, X, r, EC;

    ElGamal(String key){
        sc = new SecureRandom();
        this.key = key;
        //private key calculation
        secretKey = new BigInteger("123456789987");
        //public key calculation
        p = BigInteger.probablePrime(64, sc);
        b = new BigInteger("3");
        c = b.modPow(secretKey, p);
    }
    public BigInteger encryptKey()
    {
        X = new BigInteger(key);
        r = new BigInteger(64, sc);
        EC = X.multiply(c.modPow(r, p)).mod(p);
        brmodp = b.modPow(r, p);
        return brmodp;
    }

    public BigInteger partialDecryptKey(BigInteger brmodp){
        BigInteger crmodp = brmodp.modPow(secretKey, p);
        BigInteger d = crmodp.modInverse(p);
        BigInteger ad = d.multiply(EC).mod(p);
        return ad;
    }
}
```

FIGURE 8.10
ElGamal code (private key encryption).

Figure 8.11 shows the symmetric key algorithm.

```java
public SymmetricEncryption(String secret, int length, String algorithm)
        throws UnsupportedEncodingException, NoSuchAlgorithmException, NoSuchPaddingException {
    byte[] key = new byte[length];
    key = fixSecret(secret, length);
    this.secretKey = new SecretKeySpec(key, algorithm);
    this.cipher = Cipher.getInstance(algorithm);
}

private byte[] fixSecret(String s, int length) throws UnsupportedEncodingException {
    if (s.length() < length) {
        int missingLength = length - s.length();
        for (int i = 0; i < missingLength; i++) {
            s += " ";
        }
    }
    return s.substring(0, length).getBytes("UTF-8");
}

public void encryptFile(File f)
        throws InvalidKeyException, IOException, IllegalBlockSizeException, BadPaddingException {
    System.out.println("Encrypting file: " + f.getName());
    this.cipher.init(Cipher.ENCRYPT_MODE, this.secretKey);
    this.writeToFile(f);
}

public void decryptFile(File f)
        throws InvalidKeyException, IOException, IllegalBlockSizeException, BadPaddingException {
    System.out.println("Decrypting file: " + f.getName());
    this.cipher.init(Cipher.DECRYPT_MODE, this.secretKey);
    this.writeToFile(f);
}

public void writeToFile(File f) throws IOException, IllegalBlockSizeException, BadPaddingException {
    FileInputStream in = new FileInputStream(f);
    byte[] input = new byte[(int) f.length()];
    in.read(input);
    FileOutputStream out = new FileOutputStream(f);
    byte[] output = this.cipher.doFinal(input);
    out.write(output);
}
```

FIGURE 8.11
Symmetric key algorithm (health data encryption).

Figure 8.12 shows the encrypting health data.

```java
public void encryptData(String secretKey) {
    File dir = new File("src/input");
    File[] filelist = dir.listFiles();
    System.out.println("Private Symmetric Encryption Key : "+secretKey);
    el = new ElGamal(secretKey); // encrypt public key of symmetric encryption
    //to share key with doctor and store it
    String encrSecretKey = el.encryptKey().toString();
    randomString rs = new randomString(encrSecretKey);
    String part1 = rs.getRandomPart1();
    String part2 = rs.getRandomPart2();
    try {
        ske = new SymmetricEncryption(secretKey, 16, "AES");
        //Encryption
        Arrays.asList(filelist).forEach(file -> {
            try {
                ske.encryptFile(file);
                System.out.println("Files encrypted successfully");

            } catch (InvalidKeyException | IllegalBlockSizeException | BadPaddingException
                    | IOException e) {
                System.err.println("Couldn't encrypt " + file.getName() + ": " + e.getMessage());
            }
        });

        System.out.println("Encrypted Private key: "+encrSecretKey);
        System.out.println("Parts of patients private key : \n Social network part :"+part1+" \n Doctor's part : "+part2);

    } catch (UnsupportedEncodingException ex) {
        System.err.println("Couldn't create key: " + ex.getMessage());
    } catch (NoSuchAlgorithmException | NoSuchPaddingException e) {
        System.err.println(e.getMessage());
    }
}
```

FIGURE 8.12
Encrypting data (health).

Figure 8.13 shows the decrypting data.

```
public void decryptData(String p_key, String d_key){
    File dir = new File("src/input");
    File[] filelist = dir.listFiles();
    String secretKey = el.partialDecryptKey(new BigInteger(p_key+d_key)).toString();
    try {
        ske = new SymmetricEncryption(secretKey, 16, "AES");
        //Decryption
        Arrays.asList(filelist).forEach(file -> {
            try {
                ske.decryptFile(file);
            } catch (InvalidKeyException | IllegalBlockSizeException | BadPaddingException
                | IOException e) {
                System.err.println("Couldn't decrypt " + file.getName() + ": " + e.getMessage());
            }
        });
        System.out.println("Files decrypted successfully");
    } catch (UnsupportedEncodingException ex) {
        System.err.println("Couldn't create key: " + ex.getMessage());
    } catch (NoSuchAlgorithmException | NoSuchPaddingException e) {
        System.err.println(e.getMessage());
    } catch (Exception e){

    }

}
```

FIGURE 8.13
Decrypting data.

8.10 Conclusion

Big Data generally refers to the large amount of data generated from different sources. The Big Data security on cloud platforms is one of the most prominent research areas. Cloud computing and Big Data are the most important technologies to the modern data storage and access mechanism. Cloud computing is a large pool of systems that are connected via networks, which can be either a private network or a public network. Cloud computing provides dynamically scalable infrastructure for application and storage. Many enterprises are using cloud computing for storing and analyzing their Big Data which increases the concern of protection and security of the data stored in a cloud computing environment. This chapter has discussed the introduction to cloud computing and Big Data followed by the relationship between the two technologies. This chapter has also discussed about the need for retrieving information from widespread Big Data on the Web along with the challenges for information retrieval in Big Data and cloud. It also elaborates the classification of Big Data security by the Big Data working group at Cloud Security Alliance. Further, it also gives the details about the various encryption algorithms for the analysis of Big Data in the cloud environment. The chapter concluded with the different case studies along with the results of the different encryption techniques used to secure health data sharing. Existing technologies that can contribute to security and protection of data and to enable secure data sharing on cloud are discussed.

References

[1] M. Sharma, "Big data analytics challenges and solutions in cloud", *American Journal of Engineering Research (AJER)* 6(4), 46–51.

[2] C. M. White, "Data communications and computer networks: A business user's approach", *Thomson Course Technology*, 2009.

[3] A. Katal, M. Wazid, and R. H. Goudar, "Big data: Issues, challenges, tools and Good practices", *Sixth International Conference on Contemporary Computing (IC3)*, 2013.

[4] F. F. Ahmed, "Comparative analysis for cloud based e-learning", *Procedia Computer Science*, 65, 368–376, 2015.

[5] J. R. Vacca, *Cloud Computing Security Foundations and Challenges*, CRC Press, London, 2020. https://support.rackspace.com/how-to/understanding-the-cloud-computing-stack-saas-paas-iaas/. Last Accessed on 29 September 2020.

[6] https://support.rackspace.com/how-to/understanding-the-cloud-computing-stack-saas-paas-iaas/. Last Accessed on 29 September 2020.

[7] O. Terzo, P. Ruiu, E. Bucci, and F. Xhafa, "Data as a service (Daa S) for sharing and processing of large data collections in the cloud", *Seventh International Conference on Complex, Intelligent, and Software Intensive Systems*, 2013.

[8] M. Nezhad, et al., "Outsourcing business to cloud computing services: Opportunities and challenges", *IEEE Internet Computing*, 10(4), 1–17, 2009

[9] R. Saturi, "Data as a service (Daas) in cloud computing [Data-As-A-Service in the Age of Data] data as a service Daas in cloud computing", *Global Journal of Computer Science and Technology Cloud & Distributed* 12 11, 2012.

[10] Armbrust, et al., "A view of cloud computing", *Communications of the ACM*, 53(4), 50–58. doi:10.1145/1721654.1721672, 2010

[11] P. Géczy et al., "Cloudsourcing: Managing cloud adoption", *Global Journal of Business Research*, 6(2), 57–70, 2012.

[12] U. Aslam, et al., "Open source private cloud computing", *Interdisciplinary Journal of Contemporary Research in Business.* 2(7), 399–407, 2010.

[13] K. Goda and M. Kitsuregawa, "The History of Storage Systems", *Proceedings of the IEEE*, 100, no. Special Centennial Issue, 1433–1440, 13 May 2012, doi:10.1109/JPROC.2012.2189787.

[14] W. Tian, et al. "Optimized cloud resource management and scheduling: theories and practices", *Morgan Kaufmann*, 2014.

[15] P. C. Neves, et al. "Big Data in Cloud Computing: features and issues", *International Conference on Internet of Things and Big Data*, 2016.

[16] W.-D. Zhu, et al., "Building Big Data and Analytics Solutions in the Cloud, IBM Redbooks" 2014.

[17] S. A. Ahson and M. Ilyas (eds.). *Cloud Computing and Software Services: Theory and Techniques.* CRC Press, London, 2010. https://www.maximizer.com/blog/entering-the-age-of-big-data-as-a-service/. Last Accessed on 29 September 2020.

[18] https://www.maximizer.com/blog/entering-the-age-of-big-data-as-a-service/ Last Accessed on 29 September 2020.

[19] N. Zanoon et al. "Cloud computing and big data is there a relation between the two: A study". *International Journal of Applied Engineering Research*, 12, 6970–6982, 2017. https://cloudsecurityalliance.org/media/news/csa-releasesthe-expanded-top-ten-big-data-security-privacy-challenges/. Last Accessed on 29 September 2020

[20] https://cloudsecurityalliance.org/media/news/csa-releasesthe-expanded-top-ten-big-data-security-privacy-challenges/. Last Accessed on 29 September 2020.

[21] Y. Ma et al., "A novel approach for improving security and storage efficiency on HDFS", *Procedia Computer Science*, 52, 631–635, 2015.

[22] H. Cheng, C. Rong, K. Hwang, W. Wang, and Y. Li, "Secure big data storage and sharing scheme for cloud tenants", *China Communications*, 12(6), 106–115, 2015.

9

Using Flutter to Develop a Hybrid Application of Augmented Reality

Tarushee Kumar, Sudhanshu Sharma, Anirudh Sharma, Jatin Malhotra, and Vedika Gupta

Bharati Vidyapeeth's College of Engineering, New Delhi, India

CONTENTS

9.1 Introduction 141
9.2 Literature Review 143
9.3 Implementation 145
9.4 Application Architecture and Construction 147
    9.4.1 Front-End Development 148
    9.4.2 Back-End Development 149
9.5 Results 150
9.6 Conclusion 154
Notes 154
References 155

9.1 Introduction

In today's fast paced world, immersive computing, a technology which synthesizes an artificial world on top of the real one, has changed the way people interact with the world. People often find themselves limited in several ways to express themselves, and this is where a blend of realities comes into the picture. Irrespective of time, financial resources, physical surroundings, and weather, users can create a reality of their own, either blending it with the existing one (augmented reality) or making an entirely new one up (virtual reality). Snapchat[1] and Pokémon Go[2] are some of the numerous examples from AR field which have carved their own space in our world. There are a wide variety of ways in which one can achieve them, and developers are constantly finding new and better ways to incorporate this simulation into people's lives [20]. Society itself is becoming more digitized, and several countries are gearing up to enter the development phase of their nation. In order to facilitate this, everyone is ready to enter the digital era, a platform where ordinary tasks can be achieved with extraordinary ease and comfort [18, 19]. These new advancements make it easier for the newcomers to embrace them, and they are also visually appealing to the younger generation, who will be the ones to take this wave of technology forward.

Fast and scalable software solutions are required in order to develop programs and applications for different platforms and operating systems. Native frameworks were traditionally used for development, but now, with the advent of hybrid frameworks, the need for separate code bases is no longer required, and the cost of the entire model is lowered considerably using them. Flutter is one such hybrid framework developed by Google which can render applications for iOS, Android, and Web applications from the single code base. For the expansion of AR in our lives, 3D augmented blueprints aim to deliver Furnishz, a state-of-the-art application, which puts the entire developments made till date in AR a single package and hands it to the user. The app itself will be to place scaled-down 3-D models in our immediate surroundings and allow users to resize them and move them, according to their specifications. It will include marker less placing of augmented reality, and we hope to improve the existing technology of placing objects without a marker in the spatial environment, as currently the tracking methods used are unable to track the object marked without much accuracy. We will be using core augmented reality libraries ARCore and ARKit, developed by Google and Apple, respectively, and they provide a vast array of functionalities provided by the AR chips in their respective mobile devices, which will help us in easing the task of detecting areas and placing them in the desired locations. Marked tracking was dropped in favour of marker-less tracking due to the greater ease in using the latter, as well as the removal of the extra marker which must be available on site. Objects are selected using a QR code and are then downloaded from an online-based server. The surface is then selected using the device and Furnishz then renders the device onto the surface using ARCore and ARKit, AR libraries of iOS and Android. The use of multiuser AR is among the future scope of the project, as it has yet to be implemented in the hybrid environment.

The rest of the paper has been organized in the following manner. Section 9.2 presents the literature review, while Sections 9.3 and 9.4 cover the implementation and results, respectively. Conclusion is documented in Section 9.5, and Section 9.6 contains all the references used in the paper.

9.2 Literature Review

The research conducted in AR and hybrid application development in the past few decades has been summed up below in a chronological fashion.

The new era of teaching learning [1] will basically provide clarity of concepts to the student community. Through this new way of teaching, students will enhance their understanding of a concept or process. New Era of Teaching Learning: 3D Marker Based Augmented Reality basically focuses on involving the children in understanding of the concepts in 3D environment. This project will allow students to interact physically with the virtual objects and provide tangible feedback. The old method includes concept deliverance not based on visual aids. Now through this new teaching methodology includes different things (visual aids tools) such as projector, transparent slides, and models in 2d environment. Stereo head-mounted displays [2] (HMDs) can be used for supporting the application of augmented reality but they are not capable of perfectly blending real and virtual objects. Depth in the real world can be discerned through prompts such as accommodation and divergence. Although in stereo HMDs these prompts are not connected because the virtual object is generally projected at a constant distance, divergence varies with depth. This issue can result in partial depth approximations of virtual objects in a real environment.

For a long period of time, computer vision and image analysis [3] have been used as a center to view augmented reality. Even though sensors are deployed for the location-based AR scenarios, pose estimation is required for objects. Since AR combines virtually rendered objects in real time on the real-world landscape, computer vision [4] and object recognition are widely used to create a connection for a visually enhanced experience of the user. Fiducial markers are tracked using a camera, with the markers being tracked to a 3-D model. The device requires the location of the user using the GPS module [5] and then transmits the data back to the cloud, which determines points of interest around the user and guides them to the said spots. AR can also be used to counter forgery, as it has been tested to obtain the signature of a third party in a contract [6], as well as verify the legitimacy of the said signature.

Another invention [7] encloses a system, equipment, and method for adding lighting effects to virtual objects. Light sensors in an AR system monitor's nearby lighting conditions to collect the lighting data that can be used to create or update virtual light sources. An amazing new technology was proposed involving the solution of how one visualizes a design for a building [8], modification to a building, or extension to an existing building relative to its physical surroundings. The technique involves using a mobile-generated augmented reality platform to visualize the design in spatial context of its physical surroundings. The system was also practically used to visualize extensions (to a) building on one of the University of South Australia. Traditionally, users usually interface the applications such that it can display the output on flat 2D surfaces [9] while the input is generated with 2D or 3D devices.

Recently, a research presented constraint-based automatic placement system [10], with constraints being placed only on the physical surroundings. The aim of this research work was to deliver easy placement of objects on a real landscape. Methods used for detecting collisions [11] and developing ways to prevent them have also been researched upon in AR, and computer vision (CV) algorithms are one way to go about them. 2018 was the year in which 3-D objects were rendered using printed images sans any complicated equipment [12] and was used to accelerate learning concepts.

The drive for more capable devices is inadvertently helping in ramping up the processing power of devices capable of handling AR. Heads Up Displays [13] (HUDs) are getting more common in modern cars, where they are mostly found on during navigation and for measuring speed. As an effect of the significant technological revolution, vehicles have also gone through significant changes and improvements and the HUD is an example of that. The concept of glanceability has also been explored within a vehicle HUD. Glanceability is a property which is achieved when allowing users to grasp the displayed information with a quick glance. This thesis also focuses on unifying the mobile application development process, appealing for a common platform for both major operating systems, and have also employed a solution which helps in developing and maintaining the same app. Further additional research over HUD for glanceability [14] was done in 2018. During the testing, glanceability is the key driving factor in determining which component stays, and which component is discarded. The result is a low-fidelity prototype which has some knowledge about the needs of the user. Auditory response was also recorded for the rendered objects, in order to create a more real feeling of the same. The developments done by the said project were groundbreaking and have set a benchmark for future AR products.

"A Survey of Augmented, Virtual, and Mixed Reality for Cultural Heritage" [15] surveyed the difference in virtual, mixed reality, and augmented reality from a cultural heritage perspective. They were judged on parameters like tracking the object, virtual environment modelling, interaction interfaces, and systems they work on. The comparison between different cross platforms is summarized in "React Native vs Flutter, cross-platform mobile application frameworks" [16], as it basically summarizes an all-inclusive study between the two platforms namely React Native and Flutter. Techniques required to improve the performance of the user [17] and further reduce the mental efforts by loading it with the "projector-based spatial augmented reality" predictive of the indication for the what is laying ahead have also been researched in the past few years.

Table 9.1 presents the prevalent work on application development in different sectors, arranged in a tabular manner, indicating the system, hardware platform, software platform, libraries used, framework, and application sectors.

TABLE 9.1

Prevalent Works on Application Development in Different Sectors

System	Hardware Platform	Software Platform	Libraries Used	Framework	Application Sectors
DewataAR[a]	Smartphone	Android	Vuforia	Layar and DroidAR	E-tourism
BOTTARI[b]	Smartphone and tablets	Android	KNN algorithm (ML algorithm)	LarKc platform	Location-based recommendations
iOnRoad[c]	Smartphone	Android and iOS	FastCV	Sensafe	Traffic monitoring
Argon4[d]	Platform independent	Web browser	JavaScript	Argon.js	Research
Google Translate[e]	Smartphone and tablets	Android and iOS	Flutter	Flutter	Daily usage
Neybers[f]	Platform independent	Android and iOS	JavaScript	Argon.js	Interior designing
ArQuake[g]	Toshiba-Pentium laptop running Linux	Tinmith software system	Software system	HMDs	Gaming
Ingress Prime[h]	Smartphone and tablets	Android and iOS	ingress	React Native	AR-based game

a docplayer.net/56281033-Application-of-basic-balinese-dance-using-augmented-reality-on-android.html
b sciencedirect.com/science/article/abs/pii/S157082681200073X
c ionroad-pro.en.aptoide.com/app
d app.argonjs.io/
e translate.google.co.in/
f instagram.com/neybers/?hl=en
g tinmith.net/arquake/
h play.google.com/store/apps/details?id=com.nianticproject.ingress&hl=en_IN

As seen in Table 9.1, there are prevalent works on application development in different sectors.

9.3 Implementation

The developed application was implemented in a hybrid framework known as flutter,[3] which renders applications for multiple operating systems at the same time. The implementation of AR required four libraries, as well as some special augmented reality kits created by Apple and Google, respectively, for their operating systems.

The libraries used in flutter are as follows:

1. ARKit[4] – This is developed by Apple for the implementation of augmented reality in devices supporting their software. ARKit is like an umbrella library for the handling of augmented reality rich features and makes the processing of applications much smoother compared to ARCore, as the number of devices supported are much less compared to Android.

2. ARCore[5] – This is developed by Google for the implementation of augmented reality in devices supporting their software. With this, augmented reality can be implemented in 74% of the devices (total market share of Android), but the processing power required compared to ARKit.

3. Semaphore[6] – Semaphore is a state management utility library used in flutter to isolate the workings of a component from its surroundings, which then prevents the intermixing of the various components present in the application and prevents the disruption among components.

4. Unity[7] – Unity is the most powerful 3D rendering library available in the flutter ecosystem, and it handles the 3D mapping, displaying, rendering, and moving the files on the device much easier as compared to other devices.

The application's codebase was written in dart language, a programming language built especially for the flutter framework. Instead of placing markers [1], we used QR codes to generate the space required for the rendering of the 3D model. Since 3D models were not available for all the objects, we decided to map high-resolution images on blank 3D models [2] of the same dimensions, to increase the versatility and functionality of the app. This allowed us to look beyond the current restrictions of the 3D models and make some of our own. Marker-based AR has existed for quite some time, but Anchor-based AR is a recent field, also opening the doors to multiuser AR, in which two users can simultaneously interact with the same object in real time. The whole process consists of three steps, scanning the QR code, opening the URL embedded in the QR code, and continuing scanning the code and rendering the 3D object on top of it. Flutter is compiled in dart language, which is then simultaneously used to render applications for a variety of operating systems, ranging from iOS to android, and windows applications. Essentially, ARCore is handling the tracking of objects and building an understanding of the physical environment. The inbuilt motion tracking algorithms connect dots to points of interest and track them over a period in order to develop features. Combining the movement of the dots to the readings from the device, the library computes the orientation of the position of the object, as well as its movement through space. ARCore can recognize a variety of surfaces, ranging from flat to curved, as well as the reflecting properties of it. This helps in the better rendering of the object as it can track the reflection of light from it. This seamless synthesis of tracking the relative location of the mobile device, as well as developing a native understanding of the real world, helps in outputting a real-time feed onto the users' device.

9.4 Application Architecture and Construction

Figure 9.1 describes the basic flow of the application and navigation between the different screens of the FurnishZ. The application uses the camera of the phone to scan a QR code placed in the desired position where you want to render the object. The QR code contains a link to a 3D object hosted on a file sharing website which then downloads the model rendered onto our device and is ready to be placed, as well as an anchor containing the details of the object, i.e., where the user wants to place it, other details of the model, etc. The user then must detect a suitable surface (well lit, plane surface) and then places an anchor which is then shared with the model, as it contains the place where we want to render the object. The anchor is necessary in order to place the object, as when the object goes out of scope of the camera, we still need it to stick where we have initially rendered it, as we do not want to render it again and again. Currently, only one object is supported at a time, but we hope to expand the capability in the future, and foray into the area of multiuser AR, a feature currently unsupported by hybrid applications.

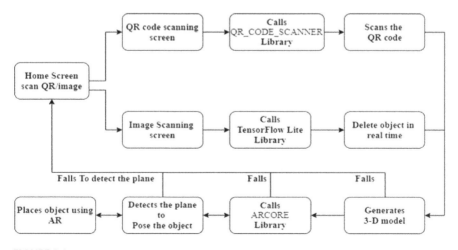

FIGURE 9.1
Flow of application.

9.4.1 Front-End Development

For the sake of simplicity, we have used the default UI layout provided by flutter for faster development. Flutter framework makes it easy for us to build user interfaces that react smoothly inside the application. This would make the transition between the application's screens smoother and improves the performance and fidelity. The whole application navigates around a single page which is a scanner menu and all the data is handled by a single provider class called MOBX. It will act as a blueprint for object creation and controls the stream of data, for example, stream of objects image capture by device's camera which is utilized by tensor flow plugin. Also, the application contains some interactive animations for the responsive user experience. UI is very minimalistic and includes some touch gestures to place an object on the detected surface. User can generate the 3D model by scanning an image or QR code with the device's camera. After the selection of the type of furniture model a user wants rendered, a 2D vector is displayed which determines the actual location of where the furniture is going to be placed. Following the rendering of the furniture, it can be viewed from multiple angles as well as be manipulated using touch inputs. It can be dragged across the screen as well as be rotated using pinch gestures.

Figures 9.2 and 9.3 are the starting pages of our mobile application which contain the scan QR button. This callback button calls the scan QR function

FIGURE 9.2
Choose from either option.

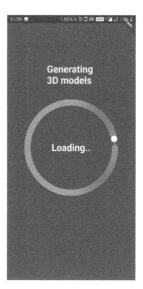

FIGURE 9.3
Loading screen of the application.

and returns the weblink of the 3D model. Now, this 3D model will be posed into the real world after it successfully detects the plane.

9.4.2 Back-End Development

Applying the Model View Controller approach, or MVC for short, allows for separation of concerns in FurnishZ. Loosely coupled UI and the back-end allow for changes to either side without affecting major changes to their interoperability. Presently, the backend is made up of a few components, ranging from interface to local storage; the local storage mainly deals with the 3D models and their storage (Figure 9.4).

FurnishZ uses a model view controller architecture as shown in Figure 9.5. Whenever the application is opened, it invokes a function responded to by the controller, which in turn works with the model to prepare the view needed by the device to render the content on the screen. An example of this would be when the user first opens the application and selects the functionality from the screen, that in turn tells the controller the functionality which is selected and helps the model to render a page for the view on the device.

Figure 9.7 shows the 3D model which is imposed into a real-world using AR generated with the help of QR code, as shown in Figure 9.6. We may need to calibrate our device to detect the plane. To avoid calibration process, make sure to have a good lighting condition and at least a distance of two meters to pose an object successfully.

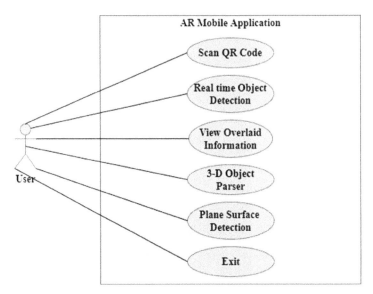

FIGURE 9.4
Use case diagram of the application.

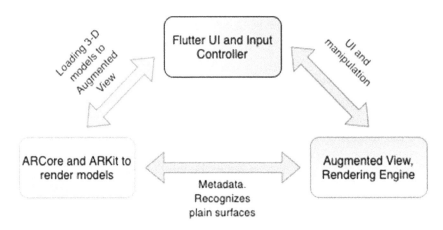

FIGURE 9.5
MVC architecture of the application.

9.5 Results

AR was implemented in the hybrid framework Flutter, developed by Google, which can render applications for iOS as well as Android Operating System from the same source code. Instead of using 3D models of objects, the application is mapping those 3D models onto a surface which then outputs a 3D

FIGURE 9.6
Sample of QR code used in the project.

FIGURE 9.7
Screenshot II: Object rendered on the plane.

model in the real world. It then detects a surface using the surface detection function inbuilt in the ARCore library of flutter and then place the object on that surface. As an additional functionality, QR codes have been utilized on the mapped area, and the object rendered on the screen matches with the QR code available online. This makes the application clean and simplistic, while also optimizing the app. The developed application can be launched and executed on two mobile operating systems, which means that the amount of resources invested in it are cut by half, and the output is doubled using the same source code. The application can place the objects in the immediate surroundings and is able to differentiate the models based on their QR code.

VisiLean[8] is a cloud-based construction management service that supports lean production planning and controlled workflow and a direct integration with Building Information Model. Table 9.2 lists out the differences between the VisiLean model and the improved model.

TABLE 9.2

Comparison of VisiLean with Furnishz

Functionalities	VisiLean model	3D-AB
3D model visualization in AR	Unable to visualize an object after placing it in the real world due to low level of information of the surroundings.	QR code contains information like depth, width, and orientation of the 3D model which not only improves visualization but also other characteristics like object textures are also found to be improved.
Time to pose the object into the real world	Rendering time was high since sometimes due to the low light and it's hard to place the object quickly.	Immediately after the scanning of the QR code, one thread downloads the data for the location from the web server and the second thread starts measuring where the user is pointing the camera of his device.
Depth sensing	Depth sensing can sometimes take a lot of time and users may have to calibrate the device several times to place an object into a real world.	The system seems to be very accurate for any target further than 2 meters from the initial location. A problem arises when the user is rotating very fast. The user experience is better when the augmented targets are further from the initial QR-code scanning location.
Multiplatform framework	We need to make separate applications for every platform which increases the overall code size.	The software uses Flutter and it is the multiplatform framework. That means we can make the application on both IOS and Android with the same code base which reduces the time and effort for the developers.
Rendering of a 3D object in low lighting.	An image is created in a camera's lens when an enough light enters the object.	The software uses amplitude-modulated (AM) light separation to take out the distorting effects of background lighting and dynamic lighting

In the VisiLean model, project performance monitor and lean construction model progress have taken a back-foot, which is a big drawback in the application itself. Moreover, visualization of performance in specific locations is still lacking in that model. Based on the review of the available technological solutions, it has been determined that an application that is able to merge Building Information Model, AR, and lean functionalities is required to maximize the improvement of construction processes on site. The above aspects clearly display the shortcomings of the VisiLean model and display how the developed model improves upon the existing model.

Table 9.3 exhibits the different render times in different lighting conditions, as well as varying distance between the surface and the device. The total rendering time of each case is the mean value of 10 outcomes (Figure 9.8).

The above graph plots the rendering speeds of the same model under different lighting conditions, thus showing that models are rendered in dimly lit conditions slower than compared to well-lit or bright environments.

TABLE 9.3

Computation Time (in ms) for Rendering 3D Model, According to Different Lighting Conditions and Distance between the Camera and Surface

	Object Distance	AR Marker Recognition	Candidate Shadow Generation	Gradient-Based Matching	Non-Maximum Suppression	Image Gradient Computation	Total Rendering Time
Direct sunlight ≈ 1,00,000 lx	2 m	3.4 ms	12.4 ms	33 ms	0.4 ms	11.0 ms	60.2 ms
Light overcast ≈ 25,000 lx	1.7 m	3.3 ms	14.7 ms	30 ms	1 ms	20 ms	69.0 ms
Open shade ≈ 10,000 lx	1.92 m	5.6 ms	11.5 ms	46 ms	3.3 ms	16 ms	82.4 ms
Near window ≈ 1,000 lx	1.3 m	8.6 ms	9.2 ms	55 ms	42 ms	15 ms	128.8 ms
Living room ≈ 500 lx	1.6 m	40 ms	2.8 ms	110 ms	166 ms	22 ms	340.8 ms
Corridors ≈ 200 lx	1.6 m	106 ms	3.2 ms	250 ms	346 ms	24 ms	529.2 ms

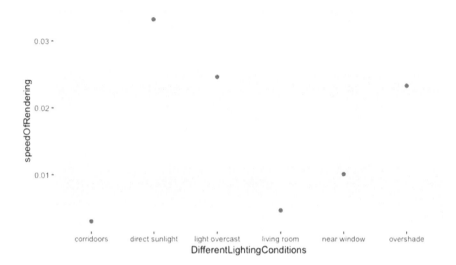

FIGURE 9.8
Graphing of different rendering speeds.

9.6 Conclusion

The rapid advancements in computer vision technology will help in further progress of AR. This is because AR can help in tackling visualization problems, a popular technique for analysis in the modern world. In this research, an idea was proposed to generate a 3D object by scanning an image and this object was placed into a real-world environment using augmented reality. The main design idea is the scanning of an image that is translated to a URL Internet site, from where the 3D object data can be downloaded. A prototype was created to test the proof of concept and demonstrate the viability of the idea. The prototype was developed on flutter framework. Flutter was specifically used so that our implementation will not be limited to a single operating system only. The developed application works best in brightly lit conditions, i.e., sunlit rooms, while dim places need optimization in the application to improve the rendering times.

Notes

1 Snapchat.com
2 Pokemongolive.com
3 Flutter.dev

4 developer.apple.com/augmented-reality/
5 developers.google.com/ar
6 pub.dev/packages/semaphore
7 pub.dev/packages/flutter_unity_widget
8 Visilean.com

References

[1] A. Klaus, A. Kramer, D. Breen, P. Chevalier, C. Crampton, E. Rose, M. Tuceryan, R. Whitaker, and D. Greer, "Distributed augmented reality for collaborative design applications", *Proceedings of Eurographics '95*, pp. C03–C14, September 1995.

[2] D. E. Breen, R. T. Whitaker, and E. Rose, "Interactive occlusion and collision of real and virtual objects in augmented reality", *Technical Report ECRC, Munich, Germany*, pp. 95–102,1995.

[3] T. Starner, S. Mann, B. Rhodes, J. Levine, J. Healey, D. Kirsch, R. Picard, and A. Pentland, "Augmented reality through wearable computing", *MIT Technical Report*, pp. 386–398, 1997.

[4] R. Raskar, G. Welch, and H. Fuchs, "Spatially augmented reality", *Proceedings of First IEEE Workshop on Augmented Reality (IWAR'98)*, pp. 63–72,1998.

[5] M. Billinghurst and H. Kato, "Collaborative augmented reality", *Communications of the ACM*, 45(7), pp. 64–70, 2002.

[6] Liang Lin, Ke Yang, and Yongtian Wang, "Research on key technology of mobile augmented reality system", *J. Chinese Journal of Image and Graphics*, 14, 560–564, 2009.

[7] T. Ewald, A. Maas, Michael R. Marner, Ross T. Smith, Bruce H. Thomas, "Quimo: A deformable material to support freeform modeling in spatial augmented reality environments", *3D User Interfaces (3DUI) 2011 IEEE Symposium*, pp. 111–112, 2011.

[8] L. Jing, "Application of augmented reality technology in teaching", *China Educational Technique & Equipment*, 35(2011), 7–8, 2011.

[9] B. H. Thomas, "Have we achieved the ultimate wearable computer?", *Wearable Computers (ISWC) 2012 16th International Symposium*, pp. 104–107, 2012.

[10] S. Ahn, H. Ko, and S. Feiner, "Webizing mobile augmented reality contents", *Proceedings of Virtual Reality 2013 (VR 2013). IEEE*, pp. 131–132, 2013.

[11] A. Jadeja, R. Mehta, and D. Sharma, "New era of teaching learning: 3D marker based augmented reality", *International Journal of Information Sciences and Techniques (IJIST)*, 6(1/2), March 2016.

[12] N. Rosa, W. Hurst, and R. Veltkamp, "Visuotactile integration for depth perception in augmented reality", *18th ACM International Conference on Multimodal Interaction*, Tokyo, Japan, 2016.

[13] M. Akçayır and G. Akçayır, "Advantages and challenges associated with augmented reality for education: A systematic review of the literature", *Educational Research Review*, 20, 1–11, 2017.

[14] P. Chen, X. Liu, W. Cheng, and R. Huang, "A review of using augmented reality in education from 2011 to 2016". In E. Popescu, K. Mohamed, K. Khribi, R. Huang, M. Jemni Nian-Shing, C. Demetrios, G. Sampson (Eds.), *Innovations in Smart Learning*. Switzerland: Springer, pp. 13–18, 2018.

[15] M. K. Bekele, R. Pierdicca, E. Frontoni, E. S. Malinverni, and J. Gain, "A survey of augmented, virtual, and mixed reality for cultural heritage". *Journal of Cultural Heritage*, 11, 1–36, 2018.

[16] W. Wu, "React native vs flutter, cross-platform mobile application frameworks". In *Information Technology*. Helsinki: Metropolia University of Applied Sciences, 2018.

[17] B. Volmer, J. Baumeister, S. Von Itzstein, I. Bornkessel-Schlesewsky, M. Schlesewsky, M. Billinghurst, and B. H. Thomas, "A comparison of predictive spatial augmented reality cues for procedural tasks", *IEEE Transactions on Visualization and Computer Graphics*, 24(11), 2846–2856, 2018.

[18] V. Gupta, and N. Rathore, "Deriving business intelligence from unstructured data", *International Journal of Information and Computation Technology*, 3(9), 971–976, 2013.

[19] V. Gupta, V. K. Singh, U. Ghose, and P. Mukhija, "A quantitative and text-based characterization of big data research". *Journal of Intelligent & Fuzzy Systems*, 36(5), 4659–4675, 2019.

[20] J. Sachdeva, M. Gupta, S. Gupta, D. Aggarwal, and V. Gupta, "Review of signboard transliteration using deep learning", *Proceedings of International Conference on Computational Intelligence, Data Science and Cloud Computing: IEM-ICDC 2020*. Springer Nature, p. 407, 2021.

10

Computational Intelligence Techniques for Recommendation System in Big Data

Avita Katal

University of Petroleum and Energy Studies, Dehradun, India

Vinayak Bajoria

InsideView Technologies, Hyderabad, India

Vitesh Sethi

University of Petroleum and Energy Studies, Dehradun, India

CONTENTS

10.1 Introduction to Big Data and Recommendation System 158
10.2 Analyzing Big Data for Recommendation System 164
  10.2.1 User Profile 165
    10.2.1.1 Profile Representation Technique 165
    10.2.1.2 Initial Profile Generation 165
    10.2.1.3 Relevance Feedback 166
    10.2.1.4 Profile Learning Technique 167
    10.2.1.5 Profile Adaption Technique 168
  10.2.2 Profile Exploitation 169
    10.2.2.1 Information Filtering Methods 169
10.3 Evaluation of Recommender System 181
  10.3.1 Predicting Ratings 181
  10.3.2 Recommending Useful Items 182
  10.3.3 Optimizing Utility 183
10.4 Book Recommendation System: An Example 184
  10.4.1 Collaborative Filtering Technique in Book
         Recommendation System 186
  10.4.2 Content-Based Filtering Technique in Book
         Recommendation System 186
  10.4.3 Hybrid Technique 187

DOI: 10.1201/9781003134138-10

10.5   A Comparison between Content-Based Filtering Techniques and
       Collaborative Filtering Techniques 188
10.6   Case Studies...188
       10.6.1   Last.fm 188
       10.6.2   MovieLens 191
10.7   Conclusion 194
References...194

10.1 Introduction to Big Data and Recommendation System

In this new era of about 8 billion people in the world, where the requirements and preferences of everyone are changing every second, it has become the need of the hour to understand what is required and design a solution so as to cater those requirements by utilizing the minimum resources.

Terabytes of data is created regularly, it may be while using social media platforms such as WhatsApp, Facebook, or simply surfing the Internet. Out of this, sore part is disposed of and some is kept safe and secured. It is really difficult to store, accumulate, and understand the trends of the data being created. This data as a whole is referred to as "The Big Data."

The big data in simpler words is basically a set of sorted information which has been sieved from the complete data created, in the form of graphs, stats, and pictures to get an idea of what actually the innovation demands. The advertisements and recommendations we get while doing online shopping, or the auto-filled search bar in Web browser, are nothing else but the big data. It helps not only in understanding trends but also in cutting short the time and the efforts required to provide consumers with the actual and optimum solution of their choice. So many MNCs and self-running organizations are competing so hard these days in order to prove their stay in the society. This puts the customers in the dominant position because now as compared to earlier times, a person has a minimum of 10 choices for a single thing they want.

By understanding the human trends and behaviours, it becomes comparatively easy for the big data developer to sort things to make the process of opting easier for the lay man. Following are the Big Data characteristics: [1]

- Volume: Volume is defined as the large amount of data generated in every second from social media, sensors, mobile phones, etc.
- Variety: Variety refers to the heterogenous nature and sources of data both unstructured and structured.

- Variability: Variability refers to the variations shown by the data, thus hampering the process being able to effectively manage and handle the data.
- Velocity: Velocity refers to the speed of the generation of data which means how fast the data is processed and generated to meet the requirements, determines actual potential in the data.
- Veracity: Veracity refers to the degree of reliability the data has to offer.

In short, these 5Vs differentiate between "data" and "The Big Data." In the twenty-first century, it can be assured that the big data has its roots spread in every field, as typical as retail market or as common as analyzing trends to win the elections. Relational Database Management System traditionally has been a standard for dealing with data. It follows the Relational model, which organizes the data into a highly structured manner.

The requirement of keeping the data in a highly structured form becomes one of the burdens when the data to be managed is in high volumes. This generally results in performance degradation as the data size goes on increasing. Thus, RDBMS fails to suffice the needs of the data called as Big Data. Therefore, a requirement of innovative and new database that can handle big data arose. The three main characteristics which prominently contribute to the necessity are as follows: volume, variety, and velocity as discussed before also.

Big Data can be calculated in petabytes, exabytes, and even zettabytes. RDBMS can achieve scaling through the growth of the servers and increasing their ability to store data. Database management systems make use of complex sharing techniques to distribute data across servers because as such they can't run on commodity hardware. Thus, scaling option used with RDBMS can be extremely costly and disruptive too. For instance, an Oracle RAC system that may hold data of any sizeable organization one day's worth will be of 20 terabytes in size which would cost millions for the organization on an RAC system. The data called Big data has given the concept of unstructured data ranging from posts extracted from various social media platforms, images, videos to time series Internet of Things data, a huge variety of data. RDBMS requires some complex technologies to fit this data into its model. Large in fact massive amount of data is being produced in real time. RDBMS doesn't support the requirements like real time working upon the data and may even suffer from down time when data may be ingested, stored, and processed.

NoSQL commonly referred as Not Only SQL is the database structure that has been introduced to tackle the needs of Big Data. It has been developed to support Big Data. They are unstructured and also meet the requirement for velocity and agility. It can store data from corner to corner on the manifold processing nodes and servers as its very foundation is upon the concepts of distributed databases. The NoSQL distributed database infrastructure is making up for few of the biggest data warehouses on the globe – i.e., the ones utilized by Google, Amazon, and the CIA.

With increasing population, the data created is drastically growing, and so is the demand of the big data creators or data handlers. This is a field which surely can explain great innovative creations in the coming future. Although NoSQL data stores are highly effective, the question of how to analyze data still goes unanswered.

Data is growing at a brisk rate; newer technologies act as hair trigger for the data production. It has been anticipated that by the year 2020 each person would be producing 1.7 megabytes of data per second [2].

Shopping in a shopping complex where clothes of all sizes are randomly kept in an unorderly fashion will waste too much of our time and create chaos. The same would be the case if there is no sorting of data done and some information on a specific topic is required. Thus, it is really important to have an efficient way to sort and arrange useful data in forms of statistical graphs, trends, etc.

These kinds of systems exist in today's world and are known as recommendation systems. Recommendation system is nothing else but a technology by means of which we can filter information that can be used to describe the human behaviour and preferences. Computationally intelligent systems use numerical data, have pattern detection capabilities, are computationally adaptable and fault tolerant, and have error rates that are close to that of humans. Computational intelligence methods, unlike conventional AI-based systems, do not necessitate the development of precise models to deal with imprecise, incomplete, and unknown data. As a result, computational intelligence methods have been extensively used in the design and execution of a broad range of intelligent systems to address challenges where formalized models are difficult to come by. Significant application areas of the computational intelligence incorporate setting mindful suggestion, customer hardware, clinical conclusion, planning of mechanical cycles, item configuration designing, mechanical technology, patent investigation, cryptology, e-learning, the travel industry proposals, and others. Each of the above application spaces utilizes specific computational intelligence technique(s) to work adequately and to satisfy the ideal objectives. The large chunks of data are called "The Big Data," and the technology or the process of sorting the Big Data in an organized manner which after processing can prove to be highly useful in understanding the data creation of figures is known as recommendation system.

Canny operators settle on choices as per the data access. Such data incorporates information about things and additionally unique profiles of different specialists on the web. Since there is so much data, a basic issue is to choose the most fitting data with which to decide. As such, a data separating technique is basic. There are three data mining approaches for making suggestions [3]:

- Collaborative filtering: Its foundation is the taste and opinion of the people. In this kind of filtering, it is assumed that people will like a thing in future too which they are liking today. The main advantage of this kind of filtering is that by using it one can easily reach to conclusions because it is easily understandable and does not use machine understandable data.

- Content-based filtering: It is based on change. This means that it believes that the preferences of people change regularly. Thus, data is collected regularly in order to reach to a latest conclusion. It uses more of machine readable form of data. The advantage of this filtering is that it creates more accurate results due to its dynamic dimension.

- Demographic filtering: It utilizes data about people to establish a connection that exists among a concerned commodity and users who jibe upon it. The systems developed taking demographic filtering as the idea suggest recommendations on the basis of the feedback it received from past clients that are within the vicinity of akin demographic traits.

The recommendation structure can experience more and more development in coming years so as to provide more filtered and useful data to the users depending on their choices and needs. This would surely satisfy the consumers to a higher extent.

There are many problems which recommendation systems face while dealing with Big Data. Some of them are enlisted as:

- Cold start – The success of the recommendation algorithms depends on the data. So, if there are no users in the system, the system may not get started [4]. This is called as the cold-start problem.

- Noisy data – Another factor that affects the success of algorithms is the data quality. If the ratings are done by people, how genuine are they or they are doing it for an incentive is also one of the major concerns.

- Privacy – Privacy is one of the major concerns in this type of applications. People generally do not want to share private information and delete the browser cache after the session. Government is also considered privacy a bigger issue thus enforcing rules around it.

- Scalability – One of the biggest issues is the scalability of algorithms having real-world datasets under the recommendation system, a huge changing data is generated by user-item interactions in the form of ratings and reviews, and consequently, scalability is a big concern for these datasets. Recommendation systems interpret results on large datasets inefficiently; some advanced large-scaled methods are required for this issue.

- Latency – It has been observed that many products are added more frequently to the database of recommendation systems but only already existing products are recommended to users as newly added products are not rated yet. So, an issue of latency arises. The collaborative filtering method and category-based approach in combination with user-item interaction can be used to deal with this issue.

- Sparsity – It happens many times when most of the users do not give ratings or reviews to the items they purchased and hence the rating model becomes very sparse which could lead to data sparsity problems; it decreases the possibilities of finding a set of users with similar ratings or interest [5].

As discussed in the above sections, due to the data scalability and many other pre-mentioned issues, an advanced approach needs to be built and followed for building recommendation systems. An open-source framework Hadoop is used for development and execution of the distributed applications used for processing of massive amounts of data. Hadoop runs on the machines in the data centres that go unexploited; such machines are known as commodity machines.

Hadoop maintains a file system made solely for it known as HDFS, Hadoop File System. Hadoop works on the principle of putting the analyzing code near the data. So, when data is uploaded to the HDFS, it partitions the data across the clusters (multiple copies are kept to handle fault tolerance) and deploys the code on to the machine which has the data on which the work was to be performed on. In HDFS, the data is organized by keys and values as in other NoSQL's databases. Each piece of data has key + unique value. If the keys have any kind of relationships among them, it is defined by the application rather than by the HDFS.

The main components that Hadoop comprises are as follows: used with a framework that can support its high level of extensibility and flexibility. These mechanisms include [6]:

- HDFS: HDFS is the distributed file system made solely for Hadoop. This file system can hold vast amount of data across the multiple nodes in a cluster based on the foundations of the distributed structure of HDFS. Hadoop comes up with a command line interface and API useful for interaction.

- MapReduce application: MapReduce is the functional programming unit in Hadoop for analyzing the data. Mapper does the data processing step and reducer is responsible for receiving the output available via mapper and then all the data applying to the same key is sorted.

- Partitioner: The main job of the partitioner component is to divide the actual data into chunks that can be worked upon by the mappers. Different partitioners are available with the Hadoop like HashPartitioner which divides the data by rows but it is also possible to design the custom partitioner.
- Combiner: The combiner class serves the purpose of minimizing the amount of data being transferred between the Map class and the Reduce class. It is placed between the Map and Reduce class. Combiner is responsible for the grouping of values with their respective keys. This is performed on a solitary node before the key-value pairs can be returned to Hadoop for appropriate execution.
- InputFormat: The usual readers do perform adequately; however, if the data format does not match the customary pre-described format, like "key, value" or "key [tab] value," then a custom input format needs to be implemented which can validate the input-specification for the job and splits up the input file (s) into logical Input Splits each of which are then assigned to an individual Mapper.
- OutputFormat: MapReduce applications can read data which is in standard InputFormat or customized InputFormat and can then be put forward via standard output formats, such as "key [tab] value" supported, but customized OutputFormats effectively interoperate with different frameworks by composing the after effect of a MapReduce work in designs comprehensible by different applications.

A Hadoop application needs to be used with a framework that can support its high level of extensibility and flexibility. These mechanisms include [7]:

- NameNode: The NameNode employs a master-slave infrastructure for distributed computation and distributed storage. It is the master of Hadoop file system that commands the slave data node daemons to perform low-level input-output tasks. The function of NameNode is input-output and memory intensive. It also acts as a bookkeeper of the Hadoop file system which means it tracks how files are converted into file blocks, which nodes stores those blocks and the status of the entire distributed file system.
- Secondary NameNode: The Secondary NameNode reviews the condition of the HDFS group and takes "previews" of the information exhibited in the NameNode. On the off chance that the NameNode

goes down, at that point the Secondary NameNode can be utilized as a reinforcement for the NameNode. This requires manual mediation; however, there is no programmed failover from the NameNode to the Secondary NameNode as yet having the Secondary NameNode will guarantee that information misfortune is ostensible. Like the NameNode, each group has a single Secondary NameNode and must be facilitated on a separate machine.

- DataNode: Each slave machine in the cluster hosts a DataNode daemon to perform the grunt work of the distributed file system – reading and writing Hadoop file system blocks to actual files on the local file system. The DataNode completes the information administration: It peruses its information blocks from the HDFS, directs the information on each physical hub, and reports back to the NameNode with information administration status.

- JobTracker: The JobTracker daemon liaison between the Hadoop and the application. The job tracker examines the execution plan by determining which files to process, monitor all tasks that are running, and assign nodes to different tasks after the submission of the code. It generally runs on the server as a master node of the cluster. The JobTracker has the ability to relaunch the tasks on a different node when the task fails.

- TaskTracker: Code execution follows the master/slave architecture just like data storage which also executes master/slave structural design. Each slave node consists of a TaskTracker daemon that is responsible for the task execution forwarded to it by the JobTracker. It is responsible to broadcast the condition of the job (and a heartbeat) to the JobTracker.

10.2 Analyzing Big Data for Recommendation System

Almost every family these days has at least one smartphone. If the number of such smartphones are used and further the data created by these smartphones is calculated, this would accumulate into a huge chunk of data, which can be termed as the Big Data.

The recommendation system is the technique which can be utilized to filter this Big Data into segments so as to provide the customer with the data of his/her desire [8]. The recommendation system has drastically changed the life of living of individuals. A simple example of this is the shopping experience; now it is so easy to sort out things as per our desire, ratings, and preferences.

A recommendation system works well when its phases are predefined logically which are as follows: data ratings, filtering, and collection [9]. The phases have been detailed below:

10.2.1 User Profile

When finding out the working of a personal system to know how it provides suggestions or estimates a user, the main question to address is user profile. The system requires as much as possible information about the user so that it can come up with satisfactory results from the starting. Therefore, the system requires a suitable method to generate the initial profile. The User profile generation and upholding needs five design parameters as follows:

10.2.1.1 Profile Representation Technique

It is the initial decision taken into account in a recommendation system as other steps depend upon this. Meticulous profiles are crucial for both content-based filtering to ensure suggestions are to the point and the collaborative filtering to ensure that clients with identical profiles are actually alike. Many techniques came under the limelight that can be used for representation of the user profiles, like purchases made earlier, navigations or surf through web or e-mails, vector of features represented through an index, an n-gram, a semantic network, an associative network, a classifier together with Bayesian networks, rating matrix, and a collection of features oriented demographically.

10.2.1.2 Initial Profile Generation

It is often pleasing to gain more and more knowledge from the user so that recommender structure can come up with satisfactory results from the beginning. However, users are often unwilling to spend their time defining his interests. Moreover, users' interests are ever changing. Thus, creation and maintenance of user profiles are entirely different aspects. Some of the techniques followed are Empty, Manual, Stereotyping, and Training Set.

(a) Empty: Some recommendation systems don't care about generating an initial profile. They begin with an empty profile. The initial profile generation phase has no meaning. In some cases, the profile can be completed via a repeated detection method whenever a new client interacts with the system.

(b) Manual: In this, a manual system is designed which asks the users to tell about their interests through keywords. This kind of system maintains transparency because if an item is being recommended to the user, then he/she knows why it's being shown. However, it holds some drawbacks. Firstly, a lot of effort is required from user's side; they are least interested in selecting keywords. Secondly, people are mostly unable to convey their interests in form of keywords which may be due to their incapability of expression or because of lack of appropriate keywords.

(c) Stereotyping: Rich in his Grundy system introduced the use of stereotypes to maintain the models of their users. In stereotyping, a primary model is created while considering the classification crisis, whose sole purpose is to produce the preliminary predictions concerning the client. Most often the data used for classification is demographic in nature like user may be asked to fill up a registration form consisting of personal details like name, age, gender, address, education, etc. This kind of data can be a challenge to acquire in the present scenario. As users no longer trust websites and withhold any personal information, even if they do provide, it can be false. This is mostly due to the lack of faith of users in websites.

(d) Training Set: Training set is pre-compilation of client communication instances used to draw conclusion about the preliminary client profile. Among the many ways to perform this, one way is that the user is requested to rate some of preestablished instances into their degree of relevancy and irrelevancy to learn about their preferences. In this case, user provides the appropriate information and the system processes it using profile learning techniques. This method is highly simplified, but someone has to choose the examples and wrong decision can lead to less precise answers.

10.2.1.3 Relevance Feedback

Generation and maintenance of user profiles by the systems require relevant information about their interests. Whenever a user interacts with computer, they cater great deal of information about themselves. Relevance feedback is the main dimension to know about users' habits, behaviour, knowledge, and interests. It has two kinds, mostly affirmative information inferring attributes liked by user and off-putting information inferring attributes he or she is not interested in. It has been claimed that negative information provides a dramatic improvement. The two most used techniques for relevance feedback are implicit in which the system automatically infers the user preferences such as navigation history, links, and time spent and savings. Explicit in which users are required to explicitly evaluate items, one such method is ratings. They exhibit what a user thinks about a product. The feelings of a user about an item can be reflected through the actions such as likes, addition to the shopping cart, purchase, or mere window shopping through clicks.

Recommendation systems can assign implicit ratings based on user activities performed. The maximum rating is 5. For an instance, regular purchasing can be assigned a rating of 5, random purchasing can get 4, likes and comments can get 3, and so on. Recommendation systems can also take into account feedback that users provide.

10.2.1.4 Profile Learning Technique

Profile learning techniques clip the relevant information and structure them depending upon the demands by the representation of profile. Learning data consists of text without structure, making its pre-processing a necessary task so that data obtained can provide relevant information in a structured format. The most commonly used techniques are Unnecessary, Term-frequency-Inverse Document frequency (TF-IDF) and feature selection (stemming, stop words), and clustering.

(a) *Unnecessary*: In this system, information is directly taken from the client in a particular format known as client profile, thus eliminating the need for preliminarily learning's about the user. Some systems which don't require profile learning techniques are as follows:

- System acquiring information from the company's database. Many e-commerce companies maintain a list of purchases against a user profile in their database. This information can be directly used.
- Collaborative filtering systems, keep a matrix of user-item ratings for each of the profiles maintained. Systems which create the initial profile through stereotyping and do not adapt it with time. This is a case of demographic filtering.

(b) *Information Indexing and Feature Selection*: Information generally doesn't contain a predefined structure thus requires a pre-processing task to be performed in order to extract structured relevant information. This can be conducted with two possible methods; first, attribute selection and second, indexing the information.

In attribute selection, we use various techniques to decrease the number of words. Some of the techniques are stop words, pruning, and stemming.

In order to index the information, we make use of the frequency of words so that an estimate about the relevancy of the documents can be made. TF-IDF can be the one of the most suitable technique that can be used for this purpose. Every file can be viewed as a vector of its weighted terms D_j:

$$D_j = \{w_{1j}, w_{2j}, w_{3j}, w_{4j} \ldots w_{nj}\}$$
$$w_{nj} = tf_{nj} * idf_n$$

(10.1)

$$idf_n = \log\left(N / df_n \right) / \log 2 \qquad (10.2)$$

Where tf_{nj} symbolizes the total counts of the nth word happening in the file D_j, idf_n represents the inverse document frequency of nth word and df_n represents the number of files in which the nth term occurs.

(c) *Clustering*: Clustering is a reference to the task of grouping similar users together on the basis of the relevant knowledge into clusters. Further, the discovered clusters are matched with the actual data to discover if there is a degree of interestingness. Conventional collaborative filtering is usually based upon matching the present client/ user profiles in opposition to the similar profiles collected by the system over time.

10.2.1.5 Profile Adaption Technique

Most of the recommender systems characteristically engage in communication for extensive period of times; client interests certainly evolve with time. This assumption requires the system to gather current relevant observations through a technique known as relevant feedback; it represents the client's current interests. Therefore, a technique is needed that also evolves with the client profile as per the new interest and let go off the old ones. Numerous techniques come within reach and have been proposed, such as manual, time window, ageing instances, combination of undersized word and a lengthy word models, a gradual forgetting function, etc.

(a) *Manual*: With certain systems, it has been observed that the client has to update the profile whenever he or she may be interested in doing so. Like the initial profile generation technique, it suffers from two drawbacks. Firstly, it takes a reasonable amount of endeavour from the user's side. Secondly, users are not able to convey their interests inform of keywords in most cases. Thus, manually updating can be complex to perform when necessities vary rapidly.

(b) *Forgetting Function*: Forgetting about things is a natural tendency in humans. Therefore, many authors have used this forgetting function to produce a weight for each surveillance depending upon its location in the space-time fabric.

Li Taoying et al. [10] have proposed a combined recommendation algorithm based on forgetting curve and improved similarity. Firstly, the improvement in Pearson similarity is done by weighted factors to improve the quality of the person similarity for high sparse data. Secondly, the Ebbinghaus forgetting curve is introduced to keep a check on the shifting of the user's interest. User score is weighted on the basis of the remaining memory of the forgetting function. The interest of the user is changing with time and is tracked by scoring that increases the user satisfaction and the accuracy of the recommendation algorithm.

10.2.2 Profile Exploitation

For recommendation of goods or actions to a user, a clever recommendation system is responsible for making decisions in accordance to knowledge in hand such as goods, profiles of other users/clients, and many more. So, it is fundamental to decide on the majority fitting information by using which conclusions can be made. Methods that filter out information are made upon the client content-based filtering techniques and collaborative filtering techniques.

10.2.2.1 Information Filtering Methods

The most relevant of all the information filtering methods are as follows: demographic, content-based, and collaborative filtering. They are analyzed deeply below. These filtering techniques have been discussed briefly in the previous sections. The detailed look of these techniques is as follows:

10.2.2.1.1 Demographic filtering

Demographic filtering uses people's descriptions to become skilled in the association of a single product and the kind of population who goes for it. The client profiles are constructed by the classification of clients in orthodox descriptions like geographical data, physiographic data (lifestyle), etc., thus representing a feature of class of users. The user's personal data is important and this can be used for the classification of the users in terms of this demographic data. The personal data of the user can be collected by user through a registration form.

Zhao Xin et al. [11] have proposed METIS, a product recommender system that detects the purpose of the purchase of the user from their microblogs in current scenario and recommend the product on the basis of matching the demographic information of the user that is taken from their public profiles with product demographics learned from microblogs and online reviews. The properties extracted from profile in microblogs and the demographics of the product extracted from the reviews of online products and microblogs are put into the learning to rank algorithms for the recommendation of the product.

Safoury Laila et al. [12] have suggested to use a new user demographic data to give recommendations instead of using the history of ratings to overcome cold start problem. Safoury Laila et al. have presented a technique for checking the utilization of various demographic attributes, such as occupation, age, and gender, for the generation of recommendation.

The two shortcomings found within the demographic filtering system are as follows:

- Demographic filtering has its roots in the abstraction of the client's interests; therefore, the structure tends to recommend similar products to people with comparable geographic profiles. These recommendations are highly generalized as every client is unique in itself.
- It doesn't support profile adaption; the user's interests are ever changing and an individual adaption to interests' changes is necessary to be incorporated.

Nevertheless, it can be useful when combined with different techniques.

10.2.2.1.2 Content-Based Filtering

Approach followed by the content-based filtering technique is to suggest items to the client based on the imagery of the previously liked and bought items. In a simple language, it can be said that they suggest products based on the similarities of items the user had previously liked. Client profiles are fashioned using attributes and features extracted from these products, assuming each client acts autonomously.

The profile is a simplification of the entered data which is often in appearance of samples of the client preferences in a given field. Products that match the client's taste are then found through prediction. Various techniques are worked upon by systems to go with the client profile and the latest products to make a decision on their degree of relevancy for the client.

Son Jieum et al. [13] have designed a content-based filtering method that utilizes a multi attribute network to effectively reflect several attributes when calculating correlations to recommend items to the users. They have measured the similarities between the directly and indirectly linked items in the analysis of the network. The proposed method employs centrality and clustering methods to consider the mutual relationships between items, as well as determine the structural patterns of these interactions.

Ali SM et al. [14] have presented a model of the recommendation that utilizes content-based filtering and genomic tags to recommend similar movies. The proposed model uses Pearson and principal component analysis to reduce the computation complexity by reducing the tags that are redundant and show less proportion of variance.

Joanna Cristry Pattey et al. [15] have proposed a framework for the recommendation system for the purchase of cosmetics. The content-based filtering method is used. Content-based filtering is done on the basis of general techniques, item profiles, and profile of users.

A technique which goes only for collaborative filtering has certain shortcomings:

- In content-based filtering, information is extracted from various sources or manually introduced, it has to be taken in notice that this information is objective in nature (e.g., Product databases). However, if an item is being selected, then it is solely upon the subjective information about the item. Therefore, attributes with greater influence are mostly not taken into account.

- Another drawback which exists is overgeneralization. This type of filtering lacks an inherent method to generate unforeseen findings. The structure has been found to recommend more or less what user had already seen or indicated a liking for. This restricts the user to see items similar to what already has been rated.

- With unadulterated content-based approach, the only criterion influencing the upcoming performance is its own feedback. Feedback doesn't show up frequently, provided the unwillingness of clients to carry out proceedings that are not heading towards their instantaneous goals, and the short communication span of the client with the system. This makes the recommendations to be imprecise in their quality.

These drawbacks can be removed by the combination of collaborative filtering with content-based filtering approach.

The various techniques for application of content-based filtering are as follows:

Content-based filtering techniques create a direct relation between new items and user in order to recommend them one. Consequently, a client-profile matching procedure is required. Many methods have been premeditated and mentioned below, which aim to computerize the progression of classifying products as relevant or irrelevant through comparison of client's interests and preferences with those of the representations of items.

(a) *Keyword Matching*: It comprises an approach that counts the words which are at hand in both the existing picture of the fresh product and in the client profile. Although this replica can have tribulations when considering synonyms and homonyms, these problems can be solved using stemming that reduces the inflectional word and derivational word form to a common base form.

Vinaya et al. [16] have proposed a keyword-based recommendation system. The aim of the proposed technique is presenting a personalized list of recommended items to the users effectively. Specifically, keywords are used to identify the preference of the user, and a user-based collaborative filtering algorithm is adopted to generate the correct recommendations. The scalability and efficiency of the proposed system is improved by implementing on Hadoop in Big Data environment.

(b) *Cosine Similarity*: Cosine similarity is derived from the information retrieval models especially vector space model which extensively uses it. It is additionally utilized as a part of frameworks with basic client profile portrayal. The client profile is treated as an index and new product/thing as query with n-dimensional vectors. The cosine spot item equation ascertains the cosine point amid the two vectors. As the value reaches "1," the two vectors coincide. On the off chance that two vectors are absolutely irrelevant, they will be orthogonal and the estimation of the cosine is "0."

$$sim(u,v) = \left(\Sigma i \ u_i.v_i \right) / \sqrt{\Sigma(u_i)^2} \sqrt{\Sigma(v_i)^2} \qquad (10.3)$$

where u is the index and v is the new item.

(c) *Nearest Neighbour*: Closest neighbour calculations depend on registering the separation from the intrigued product to either whatever remains of products or the classes of product in a client profile. This sort of calculation works by relegating things in the preparation set to the client profile. To learn the interests of the items, the product it is allotted to the nearest class.

A particular case of nearest neighbour algorithm is case-based reasoning (CBR), broadly construed, is the procedure of solving new problems based on the precedent experiences with the comparable kind of problems. In content-based filtering, client profiles are symbolized through a compilation of precedent occurrences, and to recommend fresh products, a wide range of resemblance dealings can be practical. The resemblance measure applied is supposed to be capable of encoding the knowledge of client's interests. If the cases are simple one algorithm that would work just fine is k- nearest neighbour for a given number k. So, for a given item, it is matched to k user profiles to which it is most suitable and is used to predict to which user profile it can also be recommended. Thus, based on item representation, the function could be simple keyword matching or a weighted comparison. Like weighted Euclidean distance where a particular dimension of the item can be given more importance than others by using weight term "w."

$$d(u,v) = \sqrt{\Sigma_i w_i \times \left(val(u,X_i) - val(v,X_i) \right)^2} \qquad (10.4)$$

Let "w_i" be a non-negative term that specifies a feature X_i of the item and u, v be the user profile and new item, respectively.

The recovery and adaption modus operandi from CBR are quite admirable for development of intelligent recommendation structures.

(d) *Classification*: Frameworks in view of content-based filtering can deal with the suggestion undertaking as a grouping assignment. In light of the properties of the products, the framework will attempt to incite a model for every client which enables him or her to characterize unseen products into at least two classes. A portion of the normally utilized classes are intriguing, not fascinating, applicable, indistinct, not pertinent, and so forth. This implies the client profile is represented as a classifier:

- Neural networks: Neural networks generally have an input, secondly hidden layer and finally an output layer. The input layer consists of attributes of the item and the output layer will predict whether the item is relevant or irrelevant to a particular user profile.

- Decision tree: In this, attributes of the items are root nodes and intermediate nodes, whereas the leaf nodes are define relevancy of the item to the user profile. This is easier to be understood by a layman.

- Bayesian networks: These networks are used to predict the degree of relevancy of the products for a particular client. This degree is calculated using the Bayes' theorem as below:

$$P(C_i| X) = P(X| C_i) * P(C_i) / P(X) \tag{10.5}$$

$$\text{Where } P(X| C_i) = k = 1 \prod n P(x_k| C_i) \tag{10.6}$$

where x_k represents the attributes of the item and C_i the classes of relevancy.

Kejin Cui et al. [17] have proposed a content-based recommendation system and applied Bayesian networks and ontologies into the vocabulary recommendation process for spatiotemporal data discovery. They have proposed a new method for spatiotemporal data discovery on the basis of Bayesian networks and ontologies.

Huang Z et al. [18] have proposed (TRec), an efficient recommendation system for searching passengers with deep neural structures. The presented system is based on the wide & deep model, which trained wide linear frameworks and deep neural networks together and have the advantages of memorization and generalization to search the passengers. In order to increase the accuracy of searched passengers, the presented recommendation system utilizes taxi drivers that have experience as learning objects, while considering the prediction of hunting passengers, the prediction of the condition of road, and the calculation of earnings simultaneously.

Gershman Amir et al. [19] have proposed a recommender system based on decision tree. First, the decision tree shows the list of items that are recommended at the leaf nodes, rather than single items. This leads to less searching, when utilizing the tree to compile a recommendation list for a user and consequently enables a scaling of the recommendation system. Second is the splitting method for the construction of the decision tree. Splitting is based on the least probable intersection size that is a new criterion. The new criterion computes the probability for getting the intersection for each potential split in a random split and selects the split that generates the least probable size of intersection.

10.2.2.1.3 Collaborative Filtering

The collaborative filtering matches clients with related interests or preferences and then provides them with suggestions. Recommendations to be made are found through factual investigation of the information, which prompt learning disclosure about examples and analogies of information extricated unequivocally from assessments of things given by various clients or certainly by checking the conduct of various clients in the system.

Here the products are suggested on the basis of other client's fondness rather than the previous choices made by the client. Instead of computing similarity among items, similarity between users is taken in account. In this type of filtering, a client profile only consists of the data client has provided the system with. These records are weighted against the other client's records, to find the overlaps which are then grouped together. This is then used for suggesting new products. Usually, for every client, a set of "nearest neighbours" is defined using the correlation between the precedent provided feedbacks, like ratings.

N. Divya et al. [20] have presented a framework of the recommendation system based on collaborative system using rating prediction. The authors have proposed two methods in the recommender system that prevent failure of ensuring real-time requirement of the recommendation system. The user recommendation is accepted only after the authentication of the transaction ID and one-time password (OTP). The feedback of the user is processed using support vector machine. The system only allows feedback from the authenticated persons. The product reputation is inferred from the group of persons who purchase the same product and recommendation is done on the basis of collaborative filtering.

N. Mittal et al. [21] have proposed a framework for the recommendation system application of data partitioning/clustering algorithm on ratings dataset followed by collaborative filtering to design a movie recommender system. The presented technique minimizes the time of processing and increases the prediction accuracy.

Meghna Khatri [22] have proposed a decision support framework to help companies related to ecommerce select the best collaborative filtering (CF) algorithms for generating recommendations-based online binary purchase data. An experimental design applies several CF configurations to create the presented framework, that are characterized by different data-reduction techniques, CF methods, and similarity measures, to binary purchase data sets with distinct input data characteristics, i.e., sparsity level, purchase distribution, and item–user ratio.

The important points that support the collaborative filtering are as follows: more and more people must take part to increase the likelihood that any one person will get other users with common choices, there must be any easy technique to represent the interest of the users, and the algorithms should be able to match people with common choices [23].

The above three criterions aren't easy to be formulated and hence lead to certain drawbacks, although these drawbacks can be solved via efficient use of Big Data. The drawbacks are as follows:

- Early-rater problem: At whatever point a new item shows up in the database, it is extremely unlikely that it can be recommended to different clients until and unless more data is acquired upon it by means of different clients. The data can be any type of relevant feedback like ratings.

- The sparsity problem: This issue happens when there are less number of clients in the framework in respect to the data in the system. This makes the risk of evaluations to wind up noticeably inadequate, in this way diminishing the gathering of recommendable items. Likewise, sparsity creates a genuine figuring challenge as it ends up noticeably difficult to discover neighbours and considerably difficult to prescribe things since excessively few individuals have given evaluations/ratings.

- Unusual tastes: There are constantly a few clients whose tastes differ from the standard. Thus, lion's share of the clients won't impart his insights or different preferences, prompting poor proposals.

- The trouble of accomplishing a minimum number of members makes collaborative separating tests costly. The system requires substantial measure of information from large number of sources before getting to be noticeably effective.

- Herlocker likewise presented the issue of absence of straightforwardness in the collaborative filtering. These systems today resemble secret elements, mechanized for counselling and recommendations but however can't be questioned. A client is given no flag to consult in order to choose, when to believe the recommendation system and when not.

These drawbacks could be removed by uniting the content-based filtering techniques with the collaborative filtering approach.

The various techniques for application of collaborative filtering are as follows:

Frameworks based on collaborative filtering strategies coordinate individuals with comparable interests and inclinations and after that make suggestions on this premise. It suggests things on the premise that individuals who have concurred back on certain things are well on the way to concur in future as well. As a rule, the way towards processing a suggestion comprise of three essential strides as follows:

(a) *Find Similar Users*: Typical similarity measuring techniques like nearest neighbours, clustering, and classifiers could be used to calculate the resemblance between the clients. The expanse of the current client to that with other clients is computed.

Since Big Data is being considered, measurable testing strategies should be utilized to locate an agent subset for which likeness is processed. As a rule, systems can't work with expansive arrangements of information containing every one of the clients and highlights, since the execution of the system will bit by bit separate. So, the need to diminish the information dimensionality comes to the forefront. Dimensionality reduction can be done with some of the techniques listed below:

Principle Component Analysis (PCA): It is a statistical technique that orthogonally alters the original n coordinates of the data set into a new-fangled set of n coordinates called principal components. It is an orthogonal straight change that changes information to new organized framework with the end goal that the best difference by some projection of information (variance) comes to lie on the first coordinate, the second most noteworthy variance on second coordinate and so on.

The authors in [24] have proposed an efficient Multi-Criteria Collaborative Filtering (MC-CF) algorithm using dimensionality reduction technique to improve the recommendation quality and prediction accuracy. Dimensionality reduction techniques such as principal component analysis and singular value decomposition are used to solve the scalability and alleviate the sparsity problems in overall rating. Their proposed technique implemented using Apache Mahout, which allows processing of huge dataset that stored in distributed/non-distributed file system. Some of the techniques to find similar users are as follows:

Nearest Neighbours: Nearest neighbour calculations are appropriate both for content-based and collaborative filtering. In the latter case, nearest neighbour calculations are worried in figuring the separation between clients based on their inclination history. Nearest neighbour calculations can fuse the updated information quickly, yet the scan for neighbours can be time consuming in huge databases.

In broad spectrum, the techniques mostly followed in the recommendation systems to find similarity between the clients are as follows: cosine similarity and correlation. Cosine similarity is applied in the similar way as discussed in content-based filtering.

For correlation, there are two most widely used techniques Spearman's rank correlation and Pearson correlation coefficient. Pearson correlation coefficient is derived from linear regression model and has certain set of assumptions like the relationship and errors must be linear and have probability distribution of mean 0 and unvarying variance. When these model constraints are not satisfied, it may result in false results. Pearson correlation coefficient between two users can be derived by the equation 10.7 which tries to find correlation between two users on the basis of the ratings given by both to particular item k.

$$r(X,Y) = \Sigma k (X_k - X')(Y_k - Y') / \sqrt{\Sigma k (X_k - X')^2 (Y_k - Y')^2} \qquad (10.7)$$

The ratings of a person X and Y for the item k are given by X_k and Y_k, where X' and Y' represent the mean values of their rating.

Clustering: In the mid-1970s, the client demonstrating group went for a generalization approach called as supervised learning. In this, while the system is in development stage, client subgroups are recognized and run of the mill properties of the individuals from the client subgroups is determined. Amid the run-time of the system, the client is allotted to at least one of these predefined client sub-groups. This pre-definition of these generalizations is an apparent disadvantage. As an option, the Doppelganger system utilized clustering system to discover client bunches powerfully, based on all accessible client models. Once the groups are made, expectations for an individual can be made by averaging the suppositions of alternate clients in that bunch. Clustering strategies deliver fewer individual suggestions than different techniques. They can have lesser exactness than the nearest neighbour calculations.

A portion of the clustering procedures utilized as a part of collaborative filtering are as follows:

Average Link Clustering: It is one of the fundamental clustering calculations. It is a compromise between the sensitivity of complete-link clustering towards the outliers and tendency of single-link clustering to form long chains that are not corresponding to the intuitive notion of clusters as small, spherical objects. Average-link clustering combines with each iteration, the pair of clusters with the highest cohesion. If the data points are represented as normalized vectors in a Euclidean space, we can define the cohesion G of a cluster C as the average dot product:

$$G(C) = 1/\left[n(n-1)\right]\left(gamma(C) - n\right) \tag{10.8}$$

where $n=!C!$, $gamma(C) = $ sum(v in C) sum(w in C) $<v,w>$ and $<v,w>$ is the dot product of v and w.

Spectral Clustering: Subspace clustering techniques attempt to discover clusters in the subspaces of the original data space. In certain cases, it is more effective to design a new space as an alternative of using subspaces of the original data. This type of clustering can be used for Big Data, which is known to have high number of dimensions and such dimensionality reduction methods can prove useful. In this method, firstly the affinity matrix is created and k leading eigenvectors of the affinity matrix are calculated. A clustering algorithm like k-means can be applied on the new space and finally the formed clusters are projected back to the original data.

Alternating Least Square: It has been observed that although explicit data like five-star rating scale is no easier to comprehend, data found is usually implicit in nature. This implicit data comes with a problem of missing values, which are often filled with negative values but this doesn't signify a negative behaviour.

The alternating minimum square technique handles both of these issues elegantly by taking in a factorized grid approach by methods for various certainty levels on binary preferences: unnoticed things/items are incited as negative values with low confidence, where present things/items are treated with positive values with considerably higher certainty.

Then learn user factors X_u and artists factors Y_i by minimizing confidence weighted sum of squared errors loss function:

$$Loss = \Sigma u \, \Sigma i \, C_{ui} \left(P_{ui} - X_u Y_i\right)^2 + l\left(\|X_u\|^2 + \|Y_i\|^2\right) \tag{10.9}$$

where C_{ui} is the confidence that user has positively rated an item, P_{ui} is a binary value indicating if user considered the item or not, and l is the basic $L2$ regularizer to reduce overfitting.

To minimize the user factors, we fix the item factors constant and then take derivative of loss function to calculate Xu directly:

$$X_u = \left(Y^T C_u Y + \lambda I\right)^{-1} \left(Y^T C_u P_u\right) \tag{10.10}$$

The product factors are premeditated in the alike fashion and the whole work is to reduce by harmonically moving back and forth until it congregates. Since P_u is sparse, $Y^T C_u P_u$ can be easily calculated. To calculate $Y^T C_u P_u$, they note that its equal to $Y^T Y + Y^T (C_u - 1) Y$. Negative term confidence is 1, $(C_u - 1)$ is sparse, and $Y^T Y$ can be pre-calculated for all users.

Kohonen networks: These symbolize a classification self-sorting out map (SOM), which itself is a unique class of neural systems. It changes over the high-dimensional Big Data into low-dimensional discrete map. SOM's structure yields nodes into a group of nodes, where the nodes in nearer closeness are close to each other than to different nodes, more remote separated. It takes a shot at the standards of completion, cooperation, and adaption. For each information client profile x, following three stages are performed:

Competition: For each yield hub j, compute $D(w_j, x_n)$ of the scoring function. For instance, Euclidean separation,

$$D(w_j, xn) = \sqrt{\sum (x_{ij} - x_{ni})^2} \tag{10.11}$$

where $x_n = x_{n1}, x_{n2}, x_{n3}...x_{nm}$ represents nth input vector with m field values and $w_j = w_1, w_2, w_3, ..., w_m$ be the set of m weights for a particular output node j. Find the winning node J that minimizes $D(w_j, x_n)$ overall output nodes.

Cooperation: Identify all output nodes j within neighborhood of J defined by the neighborhood size R. For these nodes, perform the adaption for all input record fields:

Adaption: adjusting the weights:

$$W_{ij,new} = w_{ij,current} + n(x_{ni} - w_{ij,current}) \tag{10.12}$$

where n, $0<n<1$, represents the learning rate and $w_{ij,new}$ represents the new weight after adaption and $w_{ij,current}$ represents the old weight.

Regulate the learning rate and neighbourhood size, as considered necessary.

Stop when terminating criteria are met.

(b) *Create a Neighbourhood*: At whatever point a system searches for comparable clients, it shapes the area of most comparative clients for the objective client. By and large two strategies are most generally used to decide what number of neighbours to choose: the relationship threshold procedure and the best-n-neighbours technique.

- The correlation threshold technique is used to set a flat out relationship edge, where all neighbours with an outright connection more prominent than given threshold are chosen. Setting a high edge restrains the area to contain good correlations, yet for some clients such high correlation corresponds to inaccessible, resulting in small neighbourhood with forecast coverage for huge number of items.
- The best n-neighbours technique is to choose the best-fixed number of clients. It performs well, if it doesn't bound prediction coverage.

(c) *Prediction on Basis of Selected Neighbours*: The last step is to extract recommendations from the user's neighbourhood. When neighbourhood is formed, ratings from these are combined to compute a prediction. Different techniques are used in different systems; some of them are as follows:

- Most-visit product proposal: As the name itself suggests that this recommendation investigates the area and sweeps through client's interests extracting the most, much of the time, the chosen product. The system sorts the products according to recurrence and returns the n most continuous products yet not chosen by the dynamic client for proposal.
- The association govern-based proposal: It uses rules generated previously instead of using whole populace of clients.
- Weighted average: In this technique, all the neighbours' ratings are consolidated and a weighted average of the ratings using the correlation as the weight is computed. The basic weighted average assumes that all users give ratings of approximately the same distribution.

10.2.2.1.4 Hybrid Approach

Hybrid systems tend to take advantage of both content-based and collaborative filtering by solving the shortcomings of either one using the other.

For content-based system, following problems can be solved by collaborative system:

- Lack of subjective: In a collaborative system, the group of clients can offer this category of data unequivocally. It can be outlook of a trustworthy third party that one can use for a certain item. For example, a person can go for spicy food, if suggested by a friend.
- Lack of novelty: Content-based framework can discover anything novel, constraining the scope of uses for which it would be valuable. Collaborative separating strategies like evaluations can take care of this issue by utilizing other individuals' suggestions.

- Lack of user information: The problem of lack and limited user information can be overcome by use of collaborative filtering. One could fill up the client information with other clients' experience as a foundation.

For collaborative systems, following problems can be solved by the content-based system:

- Early-rater problem: New items can be recommended based on their characteristics rather than feedback received for them.

- Unusual tastes: Recommendations to users with abnormal tastes can be made by content-based techniques rather than finding a similar user with unusual tastes.

- Quantity of members is likewise not vital in the content-based system as they don't rely on populace.

Along these lines, both content-based and collaborative filtering add to each other's viability, dodging the constraints said for every system and making a framework with benefits of both and accomplishing unwavering quality and serendipity.

10.3 Evaluation of Recommender System

A recommendation system is created with the sole reason to serve a business demand. It is planned around the business insight, thus, should satisfy the business objective. In this way, the proposed system should be tried before going onto the Web. For which it should be mapped into one of the three assignments underneath recommendation, utility optimization, and rating expectations. Given such a mapping, the designer should now pick which assignment metric to utilize remembering the ultimate objective to rank arrangement of candidate recommendation algorithms. It is critical that the metric match the endeavour, to avoid an uncalled or improper ranking of the candidates.

10.3.1 Predicting Ratings

In this method, the structure provides us with a set of forecasted ratings on the basis of which we tend to evaluate the system. Root mean square error (RMSE) is one of the most pioneer methods available. If $p_{i,j}$ is the predicted rating for user *i* over item *j*, and *vi,j* is the true rating, and $K = \{(i, j)\}$ is the set of hidden user-item ratings then the RMSE is defined as [25]:

$$RMSE = \sqrt{\frac{\Sigma_{(i,j)\in K}\left(P_{i,j}-v_{i,j}\right)^2}{\# K}} \qquad (10.13)$$

The other variants are the mean square error (which is equivalent to RMSE) and mean average error (MAE), and normalized mean average error (NMAE).

For example, suppose a user explicitly rates the items by the five-star rating scale where 1 means strong dislike and 5 means extreme likelihood, and also there are a bunch of such ratings from the past. A technique called split validation can be used, in which a subset of these ratings is taken in account like 70% and the recommendation system is built upon it. Now we use the rest of the 30% of the rating set to evaluate the performance of the recommendation system, like a rating of 4 is predicted as 3.5 so the error is 0.5. Then compute the average of the errors of the whole test set and get a final result. This shows how good the system works. MAE however does not penalize much for huge errors like 4 start rating being predicted as 1.

Along these lines, RMSE has a tendency to punish bigger mistakes more extremely than alternate measurements, while NMAE standardizes MAE by the scope of the evaluations for simplicity of contrasting blunders crosswise over spaces. RMSE is appropriate for the expectation undertaking, since it quantifies errors on all evaluations, either negative or positive.

10.3.2 Recommending Useful Items

It has been observed that users are often too lazy to rate anything. A user visits a webpage and if he or she likes the content then they might buy it or consume it else the session ends as fast as it began. Thus, the major part of data consists of implicit ratings and one can't measure the feedback because it was never given in the first place.

However, we can still use split validation techniques on such data. For example, 80% purchase data of the company is used as the training set for the recommendation system. Then for the test set, only a few purchases are submitted while the rest of them have to be predicted. Like out of ten, four items were hidden so precision can be 0%, 25%, 50%, and 100% depending upon the how many of the hidden four appeared in the recommended ten items. This accuracy is known as recall.

The objective is to recommend additional products to the client based on the present choices of the client. This is not a symmetric scenario and objective is to make better recommendations, not to discourage the occurrence of bad recommendations. The recommendations can be classified through Table 10.1

TABLE 10.1

Classification of Possible Recommendation of a Product to Client

	Recommended	Not-Recommended
Relevant	True positives (tp)	False negatives (fn)
Irrelevant	False positives (fp)	True negatives (tn)

Following can now be calculated through Table 10.1:

$$Precision = tp / (tp + fp) \tag{10.14}$$

$$Recall = tp / (tp + fn) \tag{10.15}$$

where "*tp*" represents true positive which means those items which are both retrieved and relevant for the user, "*fn*" represents false negatives which means those items which were relevant for the user but weren't recommended to him. "*fp*" represents false positives which means those items which were recommended to the user but didn't fill his interests and as result were irrelevant to him. "*tn*" represents the item which were irrelevant for the user and weren't recommended to him either.

Precision is the measure of how many items truly reflected the interests of user and shown to him too, to that of total recommendations made, whereas recall is the measure of number of products truly reflecting the interests of the user and recommended to him simultaneously, to that of items that truly represented users' interests [26].

10.3.3 Optimizing Utility

The manner in which a model interacts or presents itself to the user has a great impact on the affectivity of the recommendations made to the user. If the recommendations are not presented properly in a readable and understandable format, the whole recommendation system will fail to serve the purpose and that too at the last stage.

For instance, a book recommendation system must present the results with book cover being displayed along with title and an abstract so that it can catch users' sight and interest. A simple list type of structure with just book title would make the user leave faster than he came. The first case discussed will be a good estimate of expected utility, while in the second case, we could model the way users scan a list.

In this model, we evaluate an unbounded recommendation list, which consists of every item present in the catalogue. Given a list of such items, we make two main assumptions. Firstly, the user starts looking over the items from the starting of the list and we also assume that the item at kth position in the list has the probability of

$$1/2^{(k-1)(h-1)} \tag{10.16}$$

where "h" is the half-life parameter, specifying the location of the item in the list that has 0.5 probability of being viewed.

For binary case, we use the below formulae for computation of half-life utility score:

$$R_a = \sum_j \tfrac{1}{2}^{(idx(j)-1)(h-1)} \tag{10.17}$$

$$R = \sum_a R_a / \sum_a R_a^{\max} \tag{10.18}$$

where the summation in the main condition is over the favoured things just, $idx(j)$ is the file thing j in the suggestion rundown, and R_a max is the score of most ideal proposals for client a.

10.4 Book Recommendation System: An Example

Recommendation systems are being widely used to recommend items to the end user. Online book selling websites are competing with each other to retain readers by suggesting them with books suiting their interests. Recommendation systems are one strong tool that is being used by these industry giants to poach and retain the customers both.

Recommendation systems when developed with hybrid approach overcome shortcomings of collaborative and content-based recommendation systems both by the use of each other to solve them. For instance, early-rater problem for a new book can be solved by recommending it on the basis of its characteristics rather than feedback received. Similarly, a new user with limited user information can be recommended books on the basis of demography.

In this book recommendation system [27], collaborative filtering technique is used to gain inputs from the relational database to suggest recommendations; on the other hand, ontology helps to store user profiles. The client profile information is utilized by content-based techniques as contribution for prediction of recommendations. Use of ontology-based database helps to store data in hierarchical format and provides us with concrete consequences given on the data and the relationship that exists in the data. It simplifies the task of retrieval and querying of data and avoids parsing of log files.

This book recommendation system model makes use of hybrid recommendation technique which effectively combines collaborative filtering techniques with content-based techniques and demographic attributes to produce correct and precise recommendations. Collaborative filtering techniques make use of client-product fondness data in the format of ratings as contribution to make available a relational database, whereas content-based filtering technique utilizes product attributes to come up with suggestions. It suggests products by comparing them with profile of clients' interests. Figure 10.1 shows the book recommendation system.

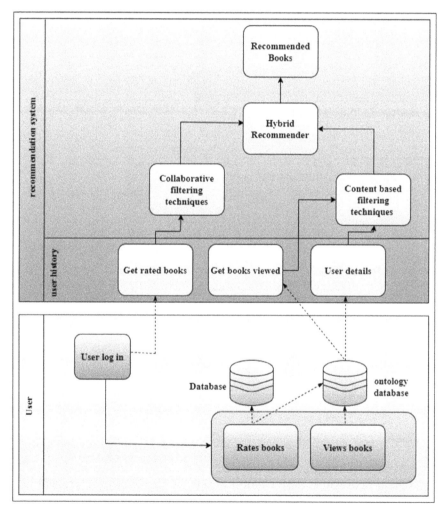

FIGURE 10.1
Book recommendation system.

10.4.1 Collaborative Filtering Technique in Book Recommendation System

Collaborative filtering technique makes use of information related to the user-item preferences which can be collected via ratings to provide with proposals. It predicts the items in light of items past ratings and evaluations given by different users that have been stored within the relational database.

Slope one calculation, which is the least complex form of non-trivial item-based collaborative filtering based on ratings, has been utilized. The calculation figures the normal contrast between each of the items and their ratings/evaluations alongside the quantity of ratings provided to each item. In the event in which a client rates numerous things, the expectations are consolidated through a weighted normal.

Algorithm: **Slope 1**

Input(s): User(s), Book(s), rating value(s)
Output: Recommended Book(s)
Begin

     Step 1:   For a given user, read book id and their corresponding rating given by the user.

     Step 2:   Calculate the rating frequency provided by the user for every book.

     Step 3:   Calculate the average distinction between the ratings provided by the community of users to that of the current users.

     Step 4:   Calculated predicted rating as the sum of the average difference in the rating and that given by the current user for whom the prediction is being made.

     Step 5:   Calculate weighted average rating for every book with respect to frequency and predicted rating.

     Step 6:   Sort the weighted ratings in the descending order.

     Step 7:   Put forward the recommendation(s) to the user.

10.4.2 Content-Based Filtering Technique in Book Recommendation System

Content-based strategies have a tendency to suggest things in light of the item/thing features and the client interests. To discover comparable books coordinating the client interests, here MinHash system is utilized which enables us to discover the similitude between the two sets. The idea that comparative things will be hashed to a same bucket can express the higher likelihood of impacts for things/items that are comparable. In this model, LSH/MinHash has been connected with the goal that likeness between the clients can be found. Comparison of the new client to that of the already registered client is found. The main objective is to recommend the books according to the likeness.

Input(s): User Profile (name, age, sex, book genre) stored in ontology database
Output: Recommended book(s)
Begin

Step 1:   Read the current user's profile

Step 2:   For making comparison read profiles of other user

Step 3:   Initialize minHash array length and i=0

Step 4:   Initialize minHash matrix to max Integer value

Step 5:   For comparison, create similarity matrix of users considered

Step 6:   Construct similarity matrix of users for comparison

Step 7:   Calculate similarity between the users.

10.4.3 Hybrid Technique

Hybrid method utilizes book proposals given by collaborative filtering system and sifts through the clients who evaluated those books. These users are again sifted through on the premise of statistic includes that are age, range, and sex. These filtered users are contrasted with discover comparability with current user utilizing content-based method. After finding the comparability esteems for users, again a channel is connected, which recommends books of categories seen by the user and appraised by the comparable users. This diminishes the quantity of comparisons of users and makes the framework compelling by giving applicable suggestions. Figure 10.1 clarifies the working of mixture book proposal procedure.

Collaborative filtering which looks outside the inclinations of individual users gives out of the container proposals. Content-based systems which channel suggestions in light of the premise of user profile alongside statistic qualities defeat the cold start issue since user preference information isn't considered. Hybrid technique consolidates the suggestions given by above systems to give more successful proposals.

TABLE 10.2

Comparison between Collaborative and Content-Based Filtering

Basis	Content-Based Filtering Techniques	Collaborative Filtering Techniques
Functionality	It recommends item to the user based on the descriptions of the previously liked and bought items.	It matches people with similar interests and preferences and then makes recommendations.
Similarity calculation	In this, similarity among the items is calculated which have been previously liked by the user, by use of classification techniques.	In this, similarity between users is calculated by the use of clustering techniques.
Essentials	A user profile must be present, with relevant information so that user profile and item matching techniques can be applied.	Many people must participate which increases the chances that a person may find another person with similar interests
Degree of automation	It is based on objective information about the items which can be automatically extracted from various sources. So, 100% automation can be achieved.	It is based on information present in the profiles of the users which have to be manually provided and updated by the user. So, 100% automation is impossible.
Nearest neighbor	Nearest neighbor algorithms are concerned with computing the distance from interested items to either the rest of products or classes of products in a client profile.	Nearest neighbor algorithms are concerned with computing the distance between the user and their preference history.
Limitations	Weighted attribute feature is not available, overspecialization and users on feedback being the only criteria influencing the future performance.	Early-rater problem, sparse data, and unusual tastes affect the performance of the system.

10.5 A Comparison between Content-Based Filtering Techniques and Collaborative Filtering Techniques

Table 10.2 shows the comparison between collaborative and content-based filtering.

10.6 Case Studies

10.6.1 Last.fm

Last.fm [28] is an Internet-based music service that allows you to listen, watch, and share to connect you to musical world. One can visualize in real time the listening habits and trends of the Last.fm's global community.

The employment of recommendation systems have given the power to suggest users with spiking artists, albums, today's most loved, around the world and many others. Figure 10.2 shows the overview of recommendation system working.

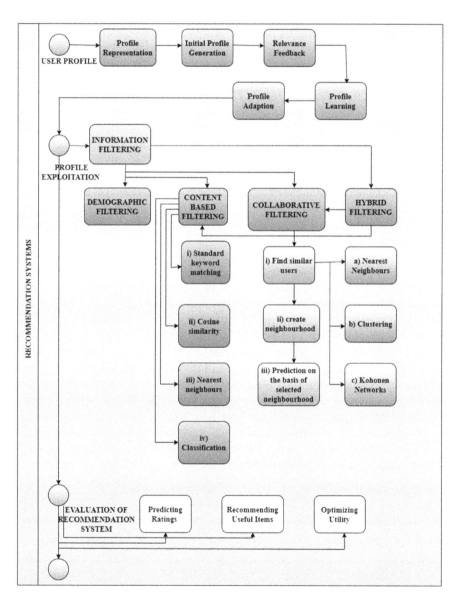

FIGURE 10.2
Overview of recommendation system working.

The suggestion framework develops a top to bottom profile of every customer's melodic inclinations, demonstrating their most loved specialists and tunes on an adjustable profile on the site, including tunes played on its station chose by methods for collaborative separating. Last.fm began as a platform for various artists to upload their music and get recognized through it. However, since most of these artists were new and so was their music, recommending their music to appropriate listeners was a great challenge. This problem could be solved by placing music into their appropriate genres. The second challenge was to find the audience for the music, which was solved by AudioScrobbler's plug-in idea. The idea was to install some plug-ins on Windows Media Player, Winamp, iTunes. These plugins submitted every song that was listened in these media players to a central database, to online personal music profile.

Last.fm's data can be visualized in the form of an undirected graph where edges run between user to user and user to artists; this can be stored in the form of an ontology database. The user-to-user relation can be of true friends with personal acquaintances or a friendship recommended by the collaborative filtering techniques by use of user-user profile matching techniques. The user to artist's relationship can be weak or strong, depending on their listening frequency or playcount.

For this situation, consider dataset of last.fm clients that was accumulated from last.fm API path in 2008. For 360000 last.fm clients, it contains top 50 most played craftsmen and in addition number of times that craftsman was played. By and large, there are around 17 million client/craftsman/playcount triples here. The tremendous volume of information here fittingly orders it as Big Data.

The least demanding approach to ascertain closeness between two craftsmen is to utilize Cosine comparability and investigate playcount as a geometry issue, by measuring the point between each combination of specialists. Every craftsman is represented to by the vector of the playcount for the 360,000 clients in the dataset. In this manner, every craftsman is represented as a vector in a 360K dimensional space formed by all users/clients. After that, the cosine angle between the craftsmen is calculated with smallest angle delineating the most similar craftsmen.

$$\text{sim}(u,v) = \left(\Sigma i \ u_i.v_i\right) / \sqrt{\Sigma(u_i)^2} \sqrt{\Sigma(v_i)^2} \qquad (10.19)$$

Where u and v are two artists and $\text{sim}(u,v)$ gives the angle between the two artists.

Cosine distance succeeds in bringing up more relevant similar artists, but there is also significantly more noise. This can be removed by

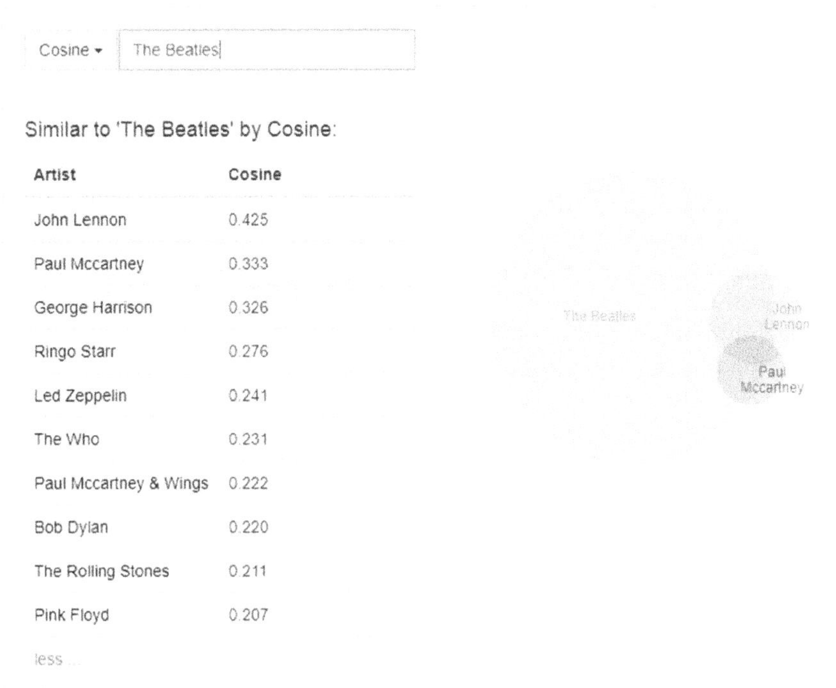

Cosine ▾ | The Beatles|

Similar to 'The Beatles' by Cosine:

Artist	Cosine
John Lennon	0.425
Paul Mccartney	0.333
George Harrison	0.326
Ringo Starr	0.276
Led Zeppelin	0.241
The Who	0.231
Paul Mccartney & Wings	0.222
Bob Dylan	0.220
The Rolling Stones	0.211
Pink Floyd	0.207

less ...

FIGURE 10.3
Artists similar to "The Beatles" that can be suggested to a user.

smoothing, by setting smoothing constant 20; the scores where there is only one user common are reduced to about 5% of their original cosine. This completely eliminated the noisy results from the head of the list here. Figure 10.3 shows the artists similar to "The Beatles" that can be suggested to a user.

10.6.2 MovieLens

MovieLens is a non-commercial, personalized movie recommendation website that helps you to find movies you will like. Users can build up an online profile through which they can share their experience, and rate movies to build up their profile as per their taste, which is then used by MovieLens to recommend user with movies to watch. The data-rich visualization provided on the site in the form of images, trailers, tags, and other expressive search tools effectively communicates the recommended results to the user.

The MovieLens Website [29] provides data collected over a period of time about its users. The particular one has 100000 ratings from 1000 users on 1700 movies. Two data files have been taken into consideration while the creation of spark code, u.data and u.item. u.data file consist of four fields userID, movieID, rating, timestamp, whereas u.item file contains fields movieID, movieName, releaseDate, imdbLink, and various genre fields. If a movie satisfies a genre, it is denoted by 1 else 0. u.data file contains about 100003 fields depicting the large volume of data and thus can be appropriately considered as Big Data, whereas u.item has 1682 fields and can be stored locally.

Information import is the initial step to be performed, which is finished by stacking the u.items record to make the mappings with results of movieID in u.data and movieName in u.item. After this, we make a start session which is open-source handling motor worked around speed and refined investigation by the utilization of its API in python. The utilization of start is important as it controls a pile of libraries including SQL, Dataframes Datasets, and MLlib for machine learning.

All the MapReduce errands are performed utilizing the start Framework. It starts by bringing in the documents from the HDFS group as a RDD (resilient distributed dataset). RDDs are permanent distributed gathering of components of information that can be put away in the memory circle over a group of machines. The information is divided among the clusters that can be worked in parallel alongside a low-level API that offers changes and activities. Each row protest is changed over to RDD.

RDD is changed over into Spark's Dataframe which is similar to RDD, an unchanging distributed gathering of information. Unlike RDD, here information is sorted out into named sections which makes the handling of these informational collections less demanding. Here, evaluations of RDD are utilized to make a start Dataframe.

Alternating least square (ALS) is a type of matrix factorization, used to train the model on ratings. ALS attempts to estimate the ratings matrix R as the product of lower rank matrices, X and Y such that $X * YT = R$. These approximations are known as factor matrices and the general approach followed here is iterative. One of the factor matrices is held constant in each iteration, while the other one is resolved using least squares. The newly solved factor matrix is held constant and the other one is resolved.

To evaluate the system, a user 0 is created with arbitrary values. Then movies which have appeared more than 100 times in the rating Dataframe are grouped together. All these movies are sorted on the basis of prediction values in the recommendations Dataframe to get the top 20 movies, which are suggested to the user 0. Figure 10.4 shows the working of the Recommendation System for movieLens.

Figure 10.5 shows the movies recommended for the user.

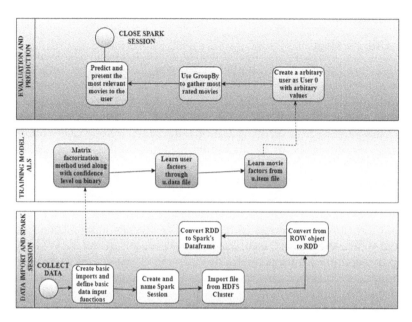

FIGURE 10.4
Working of the recommendation system for MovieLens.

```
[maria_dev@sandbox ~]$ export SPARK_MAJOR_VERSION=2
[maria_dev@sandbox ~]$ spark-submit MovieRecommendationsALS.py
SPARK_MAJOR_VERSION is set to 2, using Spark2
/usr/hdp/current/spark2-client/python/lib/pyspark.zip/pyspark/context.py:207: UserWarning:
Support for Python 2.6 is deprecated as of Spark 2.0.0
  warnings.warn("Support for Python 2.6 is deprecated as of Spark 2.0.0")

Ratings for user ID 0:
Star Wars (1977) 5.0
Empire Strikes Back, The (1980) 5.0
Gone with the Wind (1939) 1.0

Top 20 recommendations:
(u'Wrong Trousers, The (1993)', 5.749821662902832)
(u'Fifth Element, The (1997)', 5.2325282096862793)
(u'Close Shave, A (1995)', 5.0506253242492676)
(u'Monty Python and the Holy Grail (1974)', 4.9965953826904297)
(u'Star Wars (1977)', 4.9895946368408203)
(u'Army of Darkness (1993)', 4.980320930480957)
(u'Empire Strikes Back, The (1980)', 4.972929545288086)
(u'Princess Bride, The (1987)', 4.9577054977416992)
(u'Blade Runner (1982)', 4.9106745719909668)
(u'Return of the Jedi (1983)', 4.7780814170837402)
(u'Rumble in the Bronx (1995)', 4.6917591094970703)
(u'Raiders of the Lost Ark (1981)', 4.6367182731628418)
(u"Jackie Chan's First Strike (1996)", 4.632108211517334)
(u'Twelve Monkeys (1995)', 4.6148405075073242)
(u'Spawn (1997)', 4.5741710662841797)
(u'Terminator, The (1984)', 4.5611510276794434)
(u'Alien (1979)', 4.5415172576904297)
(u'Terminator 2: Judgment Day (1991)', 4.529487133026123)
(u'Usual Suspects, The (1995)', 4.5179119110107422)
(u'Mystery Science Theater 3000: The Movie (1996)', 4.5095906257629395)
[maria_dev@sandbox ~]$
```

FIGURE 10.5
Movies recommended for the user.

10.7 Conclusion

Data is being generated at a higher velocity and in higher volumes, which makes it necessary to use modern technologies and methods to handle this enormous amount of data carefully. The recommendation systems use the techniques to filter the Big Data and provide desired results/recommendations to the customer. In general, computational intelligence approaches are useful not only in context-aware recommender systems but also in context-aware scanning. In this chapter, a complete and up-to-date method for the creation of recommender system has been presented using Big Data technologies. The chapter has three main sub sections: basic understanding of recommendation system and Big Data; client profile generation; profile exploitation in which various information filtering techniques like content based, collaborative, demographic, and hybrid techniques are discussed in detail. Within each group of techniques, various methods that can help us to achieve the goals of recommendation system are also discussed. Various evaluation methods of the recommendation systems are discussed along with comparison between the content-based filtering and collaborative filtering techniques. The chapter has covered many case studies like recommendation system for book recommendations to help the reader better understand the approach to build recommendation systems. A framework for book recommendation system has been proposed, and also the chapter supplies a wide-ranging clarification of existing recommender systems with Big Data implementation.

References

[1] A. Katal, M. Wazid, and R. H. Goudar. "Big data: Issues, challenges, tools and good practices," *2013 Sixth International Conference on Contemporary Computing (IC3)*, Noida, 2013, pp. 404–409, doi:10.1109/IC3.2013.6612229. https://www.forbes.com/sites/bernardmarr/2015/09/30/big-data-20-mind-boggling-facts-everyone-must-read/#2204dd9f17b1. Last Accessed on 6 August 2020.

[2] https://www.forbes.com/sites/bernardmarr/2015/09/30/big-data-20-mind-boggling-facts-everyone-must-read/#2204dd9f17b1. Last Accessed on 6 August 2020.

[3] L. Narke and A. Nasreen. "A comprehensive review of approaches and challenges of a recommendation system", *International Journal of Research in Engineering, Science and Management* 3(4), 381–384, 2020.

[4] A. K. Pandey and D. S. Rajpoot. "Resolving Cold Start problem in recommendation system using demographic approach," *2016 International Conference on Signal Processing and Communication (ICSC)*, Noida, 2016, pp. 213–218, doi:10.1109/ICSPCom.2016.7980578.

[5] G. Guo. "Improving the performance of recommender systems by alleviating the data sparsity and cold start problems", *Proceedings of the Twenty-Third International Joint Conference on Artificial Intelligence*, 2013.

[6] R. Velusamy, K. R. Rakshitha, K. Sruthi, and S. Guruprasaath. "Big data with hadoop – for data management, processing and storing", *International Research Journal of Engineering and Technology (IRJET)* 4, 1111–1116, 2017.

[7] S. Saggar, R. Saggar, and N. Khurana. "HADOOP multiple job trackers with master slave replication model", *International Journal of Advance Research in Computer Science and Management Studies*, 3(8), 135–144, 2015.

[8] J. P. Verma, B. Patel, and A. Patel. "Big data analysis: Recommendation system with Hadoop framework," *2015 IEEE International Conference on Computational Intelligence & Communication Technology*, 2015. https://www.kdnuggets.com/2015/10/big-data-recommendation-systems-change-lives.html. Last Accessed on 15 July 2020.

[9] https://www.kdnuggets.com/2015/10/big-data-recommendation-systems-change-lives.html. Last Accessed on 15 July 2020.

[10] T. Li, L. Jin, Z. Wu, and Y. Chen. "Combined recommendation algorithm based on improved similarity and forgetting curve." *Information* 10(4), 130, 2019.

[11] X. W. Zhao, Y. Guo, Y. He, H. Jiang, Y. Wu, and X. Li. "We know what you want to buy: A demographic-based system for product recommendation on microblogs," *Proceedings of the 20th ACM SIGKDD International Conference on Knowledge Discovery and Data Mining*, 2014.

[12] L. Safoury and A. Salah. "Exploiting user demographic attributes for solving cold-start problem in recommender system." *Lecture Notes on Software Engineering*, 1(3), 303–307, 2013.

[13] J. Son and S. B. Kim. "Content-based filtering for recommendation systems using multiattribute networks." *Expert Systems with Applications* 89, 404–412, 2017.

[14] S. M. Ali, G. K. Nayak, R. K. Lenka, and R. K. Barik. "Movie recommendation system using genome tags and content-based filtering." *Advances in Data and Information Sciences*, 38, 85–94, 2018.

[15] J. C. Patty, E. T. Kirana, and M. S. D. Khrismayanti Giri. "Recommendations system for purchase of cosmetics using content based filtering." *International Journal of Computer Engineering and Information Technology* 10(1), 1–5, January 2018.

[16] V. B. Savadekar and P.B. Gosavi. "Towards keyword based recommendation system." *International Journal of Science and Research (IJSR)* 3(11), 2014.

[17] C. Kejin, J. Yongyao, L. Yun, and P. Dieter. "A vocabulary recommendation method for spatiotemporal data discovery based on Bayesian network and ontologies." *Big Earth Data* 3(3), 220–231, 2019.

[18] Z. Huang, G. Shan, J. Cheng, and J. Sun. "TRec: An efficient recommendation system for hunting passengers with deep neural networks." *Neural Computing and Applications* 31(S1), 209–222, 2018.

[19] A. Gershman, A. Meisels, K. Lüke, L. Rokach, A. Schclar, and A. Sturm. "A decision tree based recommender system." In G. Eichler, P. Kropf, U. Lechner, P. Meesad, & H. Unger (Eds.), *International Conference on Innovative Internet Community Systems (I 2 CS) – Jubilee Edition 2010*, 170–179, 2010.

[20] N. Divya, S. Sandhiya, S. Sandhiya, and D.R. Anita Sofia Liz. "A collaborative filtering based recommender system using rating prediction", *International Journal of Pure and Applied Mathematics*, 119(10), 1–7, 2018.

[21] N. Mittal, R. Nayak, M. C. Govil, and K. C. Jain. "Recommender system framework using clustering and collaborative filtering," *2010 3rd International Conference on Emerging Trends in Engineering and Technology*, Goa, 2010, pp. 555–558, doi: 10.1109/ICETET.2010.121.

[22] S. Geuens, K. Coussement, and K. W. De Bock. "A framework for configuring collaborative filtering-based recommendations derived from purchase data." *European Journal of Operational Research* 265(1), 208–218, 2018.

[23] M. Khatri. "A study of collaborative filtering approach for temporal dynamic web data", *International Journal of Advanced Networking and Applications (IJANA)*, 4(2), 1568–1573, 2012.

[24] D. Bokde, S. Girase, and D. Mukhopadhyay. "An item-based collaborative filtering using dimensionality reduction techniques on Mahout framework." (2015). arXiv:1503.06562

[25] A. Esteban, A. Zafra, and C. Romero. "A hybrid multi-criteria approach using a genetic algorithm for recommending courses to university students." *International Conference on Educational Data Mining* (2018).

[26] Precision and recall in recommender systems. n.d. And some metrics stuff. https://medium.com/@bond.kirill.alexandrovich/precision-and-recall-in-recommender-systems-and-some-metrics-stuff-ca2ad385c5f8. Last Accessed on 24 July 2020.

[27] G. Sheetal and M. Debajyoti. "Introducing hybrid technique for optimization of book recommender system." *Procedia Computer Science* 45, 23–31, 2015. http://www.benfrederickson.com/matrix-factorization/. Last Accessed on 26 July 2020. https://movielens.org/. Last Accessed on 26 July 2020.

[28] http://www.benfrederickson.com/matrix-factorization/. Last Accessed on 26 July 2020.

[29] https://movielens.org/. Last Accessed on 26 July 2020.

11

Predicting Melanoma Tumor Size through Machine Learning Approaches

Dhruv Chadha, Nikita Jain, and Vedika Gupta
Bharati Vidyapeeth's College of Engineering, New Delhi, India

CONTENTS

11.1 Introduction 197
11.2 Methodology 200
    11.2.1 Data Collection 200
    11.2.2 Data Preprocessing and Feature Engineering 201
    11.2.3 Model Selection 203
        11.2.3.1 XGBoost 203
        11.2.3.2 LightGBM 205
        11.2.3.3 Extra Tree Regressor 206
        11.2.3.4 Bagging Regressor 206
        11.2.3.5 Catboost Regressor 206
    11.2.4 Hyperparameter Tuning 208
    11.2.5 Repeated K-Fold Cross-Validation 208
11.3 Result 210
    11.3.1 Bayesian Optimization 210
    11.3.2 Experimental Result and Comparison 214
11.4 Conclusion 216
References 216

11.1 Introduction

Melanoma, also known as malignant melanoma, is a type of skin cancer that develops from the pigment-producing cells known as melanocytes [1]. The primary cause of melanoma is ultraviolet light (UV) exposure in those with low levels of the skin pigment melanin. The UV light may be from the sun or other sources, such as tanning devices. Certain phenotypic characteristics, that is, red or blond hair, light-colored eyes, medium skin tone, and pronounced mole pattern, are key risk factors for the growth of skin melanoma [2, 3]. In women, melanoma most commonly occurs on the legs, while in men, they most commonly occur on the back. About 25% of melanomas develop from moles. Changes in a mole that can indicate melanoma include an increase in size, irregular edges, change in color, itchiness, or skin breakdown. Globally, in 2012, it recently occurred in 232,000 people. In 2015, there were 3.1 million people with active disease, which resulted in 59,800 deaths. Australia and New Zealand have the highest rates of melanoma in the world. There are also high rates in Northern Europe and North America, while it is less common in Asia, Africa, and Latin America. Melanoma is more dangerous because of its ability to spread to other organs more rapidly if it is not treated at an early stage. Treatment can completely cure melanoma in many cases, especially when it has not spread extensively. Therefore, it is necessary to predict melanoma early.

Machine learning, a well-established algorithm, is extensively used in the field of predictive analysis given a dataset with different features. Machine learning is often used to build predictive models by extracting patterns from large datasets. These models are then used in predictive data analytics applications that include price prediction, risk assessment, predicting customer behavior, and document classification. Past data can be used to predict future outcomes. Machine learning offers different models that can be applied in case of different problems. In the case of regression problems, different models such as SVR, XGBoost, LightGBM, etc., can be applied to give good results in predicting the values based on the training data that is provided. Various machine learning models need to be tried and tested before choosing the best fit for the provided data. The prediction was analyzed based on the different metrics, and the final model was chosen that gave the best results considering different metrics. The prediction can be made using the model specified with a given set of hyperparameters.

Developments in the field of machine and deep learning can help doctors in early detection as well as treatment of various serious diseases including malignant melanoma. Various researchers have contributed to the field of melanoma for diagnosis of malignant melanoma as well as detection of the severity of malignant tumors. In 2017, Esteva [4] built the first deep learning, convolutional neural network (CNN) image classifier that worked on a par with 21 board-certified dermatologists to classify images with malignant lesions [5]. Authors in [6] try to check the usefulness of ABCD (Asymmetry, Border, Color, Diameter) formula in Total Dermatology Score for diagnosis of melanoma using LERS (Learning from Examples based on Rough Sets), whereas in [7], authors devise an algorithm to access the likelihood of a benign skin lesion to transform into a malignant melanoma tumor using a nevi classifier on images. Grzymala-Busse made a melanoma prediction model using K-nearest neighbor and LEM2 algorithms in [8]. The authors in [9] reviewed 25 existing models for risk prediction for melanoma and urged for further validation of these models instead of developing further models based on this review. In [10], authors try to make an automated skin lesion classifier using experimenting with various deep neural networks like PNASNet-5-Large [11], InceptionResNetV2 [12], SENet154 [13], and InceptionV4 [12] for classifying various processed and augmented dermoscopic images reported the best validation score of 0.76 for PNAS-5-Large. They also state that further improvements can be seen by more careful hyperparameter tuning as well as training on a larger dataset could offer improvements. Through this paper, authors try to predict the size of malignant melanoma tumor based on medical data having various attributes like damage ratio, exposed area to the tumor, etc.

In [14], authors showed through logistic regression modeling that the melanoma risk associated with a given level of sun exposure during adulthood increased with higher sun exposure during childhood, but the increase in risk was higher than the simple addition of melanoma risk associated with sun exposure during childhood or adulthood. Authors in [15] evaluate the quality of reporting of prediction models using the Transparent Reporting of a Multivariable Prediction Model for Individual Prognosis or Diagnosis (TRIPOD) checklist, and it was concluded that current reporting of melanoma multivariable prediction models does not meet standards. In [16], the author used the lesional steps in tumor progression and multi-variable logistic regression to develop a prognostic model for primary, clinical stage I cutaneous melanoma. The model proposed is 89% accurate in predicting survival. A completely automated system of dermatological disease recognition through lesion images, a machine intervention in contrast to conventional medical personnel-based detection, was proposed by authors in [17]. The model proposed is designed into three phases consisting of data collection and augmentation, designing model, and finally prediction. They used multiple AI algorithms like CNN and SVM and amalgamated it with image processing tools to form a better structure. In [18–20], authors have used machine learning and deep learning algorithms for prediction of melanoma tumor. In [21], authors proposed a methodology using ensembling strategy which can be used to predict the size of the tumor. In future, we can also predict the size using deep learning approaches as described in the paper [22].

Upcoming sections discuss the methodology adopted in this paper to predict the size of malignant melanoma tumors. Then the various techniques used for the evaluation of different machine learning models as well as Results and Discussions have been shown in Section 11.4.

11.2 Methodology

Machine learning offers a method for making predictions about the size of malignant tumors. Data need to be gathered before moving further. This is very important because the quality and quantity of data that you gather will directly determine how good your predictive model can be. After that, different processes need to be followed before fitting into a model. Figure 11.1 shows the pipeline that was followed while making a prediction for the tumor size.

11.2.1 Data Collection

The dataset for tumor size was collected which has a direct impact on the size of tumor. For the prediction of size, different features need to be taken into account for the accurate prediction. Dataset was collected which contains the following features. The mass of the area understudy, the size of the area understudy, ratio of normal to the malign surface, unrecoverable area of skin damaged by the tumor, total area exposed to the tumor, standard deviation of malign skin measurements, error in malign skin measurements, penalty applied due to measurement error in the lab, and the ratio of damage to total spread on the skin. It is important to study the above-mentioned factors to predict the size of tumor. Violin plots showing distribution of data in various features of dataset are shown in Figure 11.2. These features like the ratio of normal to the malign surface tell us about the spread of melanoma; unrecoverable area of skin damaged by the tumor can tell us about the after effects and thus is a key factor through which size can be predicted. The total area exposed to the tumor tells us about the damage that the melanoma causes and thus melanoma can be further prevented. Melanoma can spread almost

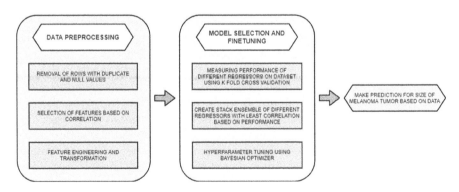

FIGURE 11.1
Flow diagram for proposed methodology.

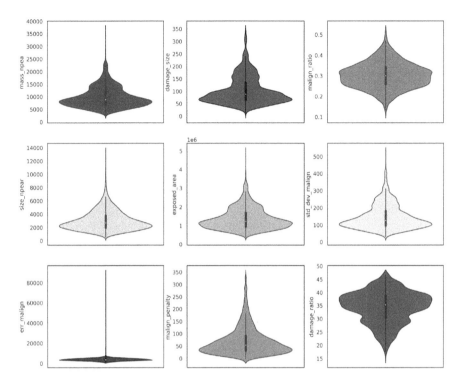

FIGURE 11.2
Violin plots showing distribution of data in various features of dataset.

anywhere in the body, but the most common sites of spread are the lungs, liver, brain, bones, and the skin or lymph nodes in other parts of the body. These features were studied and the impact on the size of tumor was noted and the following preprocessing steps need to be done.

11.2.2 Data Preprocessing and Feature Engineering

The data extracted have to be preprocessed before fitting into a suitable machine learning model to achieve better results as the data need to be in a proper manner. So, a lot of features were extracted from the dataset and then preprocessing of the data was done. The data preprocessing part included checking for any null value in the dataset because some specified machine learning algorithms like random forest do not support null values. Correlation between different features was calculated as shown in the Table 11.1 and correlation matrix is shown in Figure 11.3 and different feature engineering techniques were applied to the dataset. Feature engineering refers to creating new features from existing ones, of coming up with new

TABLE 11.1

Correlation between Different Features

	mass_npea	size_npear	malign	damage size	exposed_ area	Std_dev_ malign	err_malign	malign_ penalty	damage_ ratio	Tumor_size
mass_npea	1.000000	0.907335	0.123411	0.930691	0.998167	0.968006	0.617207	0.654849	0.901106	-0.004679
size_npear	0.907335	1.000000	0.499398	0.794096	0.903502	0.907932	0.570343	0.593019	-0.791768	0.164761
malign	0.123411	0.499398	1.000000	0.029167	0.119389	0.196430	0.081228	0.102475	-0.068605	0.374273
damage size	0.930691	0.794096	0.029167	1.000000	0.925266	0.938202	0.535834	0.676904	-0.892006	-0.163804
exposed_area	0.998167	0.903502	0.119389	0.925266	1.000000	0.962462	0.617149	0.646444	-0.900717	-0.003641
std_dev_malign	0.968006	0.907932	0.196430	0.938202	0.962462	1.000000	0.597842	0.667264	-0.884481	-0.030085
err_malign	0.617207	0.570343	0.081228	0.535834	0.617149	0.597842	1.000000	0.381454	-0.583945	0.004484
malign_penalty	0.654849	0.593019	0.102475	0.676904	0.646444	0.667264	0.381454	1.000000	-0.646893	0.011871
damage_ratio0	-0.901106	-0.791768	-0.068605	-0.892006	-0.900717	-0.884481	-0.583945	-0.646893	1.000000	0.054892
tumor_ size	-0.004679	0.164761	0.374273	-0.163804	-0.003641	-0.030085	0.004484	0.011871	0.054892	1.000000

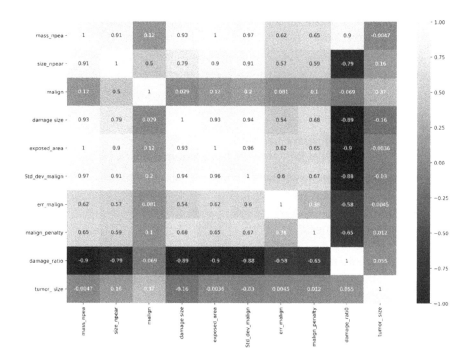

FIGURE 11.3
Correlation between different features.

variables from the list your dataset currently has and creating a new dataset which is more easily used by machine learning algorithms. Logarithmic transformation was done on some features to decrease the effect of the outliers due to the normalization of magnitude differences.

11.2.3 Model Selection

After preprocessing of data, a model needs to be selected to give the least error and get a better prediction. Different models were tried and tested on various parameters. Different models that were applied are as follows:

11.2.3.1 XGBoost

XGBoost or eXtreme Gradient Boosting [23] is an optimized distributed gradient boosting library designed to be highly efficient, flexible, and portable. It is a decision tree-based ensemble machine learning algorithm and is used for implementing machine learning algorithms under the gradient boosting framework [24, 25]. Generally, XGBoost is fast when compared to other implementations of gradient boosting. The objective function (loss function and regularization) at iteration p that we need to minimize is the following:

$$L^{(x)} = \sum_{p=1}^{n} l\left(b_p, \hat{b}_p^{(x)} + g_x(a_p)\right) + \Omega(g_x) \tag{11.1}$$

where,

$L^{(x)}$ = Regularised objective function,
b_p = Real value or label known from the training dataset,
$\hat{b}_p^{(x-1)}$ = Predicted value,
$\Omega(g_x)$ = Regularisation term,
l = Differentiable convex loss function, and
g_x = set of base learners

The regularized term is given by:

$$\Omega(g_x) = \gamma T + \frac{1}{2}\beta \sum_{k=1}^{T} \omega_k^2 \tag{11.2}$$

where,

γ is the penalty coefficient, and
$\frac{1}{2}\beta \sum_{j=1}^{T} \omega_j^2$ is the L2 norm of leaf scores.

After p iterations, the function of the model is the $(p-1)^{\text{th}}$ iteration prediction function plus a new decision tree:

$$\hat{b}_p^{(x)} = \hat{b}_p^{(x-1)} + g_x(a_p) \tag{11.3}$$

The updated objective function is as follows:

$$L^{(x)} = \sum_{p=0}^{n} l\left(b_p, \hat{b}_p^{(x-1)} + g_x(a_p)\right) + \Omega(g_x) \tag{11.4}$$

The Taylor expansion of the objective function is defined by:

$$L^{(x)} = \sum_{p=1}^{n} l\left(b_p, \hat{b}_p^{(x-1)} + f_p g_x(a_p) + \frac{1}{2}h_p g_x^2(a_p)\right) + \Omega(g_x) \tag{11.5}$$

where,

f_p is the first derivative of loss function, and
h_p is the second derivative of loss function.

Now, f_p sand h_p are defined as follows:

$$f_p = \frac{\partial}{\partial \hat{b}^{x-1}}\left(l\left(b_p, \hat{b}_p^{(x-1)}\right)\right) \tag{11.6}$$

$$h_p = \frac{\partial^2}{\partial \hat{b}^{x-1}}\left(l\left(b_p, \hat{b}_p^{(x-1)}\right)\right) \tag{11.7}$$

Here we choose the XGBoost as the regressor along with other ensemble models. Now, the hyper-parameters need to be tuned with Bayesian optimization so as to improve the performance of our final model which is described below.

11.2.3.2 LightGBM

LightGBM [26] is a fast, distributed, high-performance gradient boosting framework based on the decision tree algorithm, used for ranking, classification, and many other machine learning tasks. LightGBM can handle the large size of data and takes lower memory to run. Another reason why LightGBM is popular is because it focuses on the accuracy of results. Since it is based on decision tree algorithms, LightGBM grows trees leaf-wise (best-first). This has been shown in Figure 11.4. It will choose the leaf with max delta loss to grow. Holding leaf fixed, leaf-wise algorithms tend to achieve lower loss than level-wise algorithms. Leaf-wise may cause overfitting when data is small, so LightGBM includes the max_depth parameter to limit tree depth. However, trees still grow leaf-wise even when max_depth is specified. It results in much better accuracy which can rarely be achieved by any of the existing boosting algorithms. Fixing the max_depth of the tree also helps in avoiding overfitting.

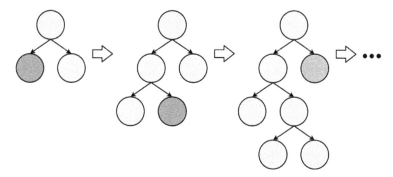

FIGURE 11.4
Leaf-wise tree growth in LGBM.

11.2.3.3 Extra Tree Regressor

Extremely Randomized Trees [27], or Extra Trees for short, is an ensemble machine learning algorithm. Specifically, it is an ensemble of decision trees and is related to other ensembles of decision trees algorithms such as bootstrap aggregation (bagging) and random forest. The Extra Trees algorithm works by creating a large number of unpruned decision trees from the training dataset. Predictions are made by averaging the prediction of the decision trees in the case of regression or using majority voting in the case of classification. The Extra Trees algorithm fits each decision tree on the whole training dataset. It can be used for classification as well as regression problems. For regression problems, predictions are made by averaging predictions from decision trees. In Extra Trees, randomness comes from random splits of data instead of bootstrapping of data as seen in random forests or decision trees. That is why it is named as Extremely Randomized Trees.

11.2.3.4 Bagging Regressor

A Bagging Regressor [28] is an ensemble meta-estimator. It trains base regressors each on random subsets of the original dataset and then aggregates their individual predictions by averaging out the values to form a final prediction as the target variable is numeric. This meta-estimator is seen as a way of reducing a black-box estimator's variance (e.g., a decision tree), integrating randomization into its construction process, and then making an ensemble out of it. This also helps to avoid overfitting of the regressor. In this meta-algorithm, various splits of data are bootstrapped and then trained on different regression models and these models are finally aggregated to produce the meta-model which is used to make predictions. That is why a Bagging Regressor is also called a Bootstrap Aggregating Regressor. The structure of a Bagged Regressor is shown in Figure 11.5.

11.2.3.5 Catboost Regressor

Catboost [29] is a machine learning algorithm that uses gradient boosting on decision trees and is developed by the Russian search engine company Yandex. It is available as an open-source library. This algorithm can provide great results without parameter tuning since it uses algorithms like ordered boosting and random permutations. Catboost uses symmetric or oblivious trees which help in decreasing the prediction time. This algorithm can be used with both big and small datasets and has an inbuilt overfitting detector which stops training when cross-validation error starts increasing.

11.2.4 Hyperparameter Tuning

Bayesian model-based approaches can find better hyperparameters in less time because they justify the right set of hyperparameters to be tested on the

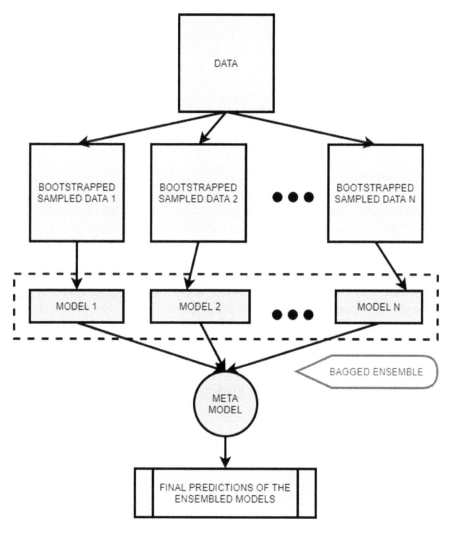

FIGURE 11.5
Mechanism of Bagging Regressor.

basis of past studies. Hyperparameter optimization is represented in equation form as:

$$y = \arg\min f(x), \text{where } x \in X \tag{11.8}$$

$f(x)$ represents an objective score to minimize such as RMSE or error rate that is to be evaluated on the validation set;
y is the set of hyperparameters that yield the lowest value of the score, and x can take on any value in the domain X.

The problem with hyperparameter optimization is that it is extremely expensive to test the objective function of finding a score. With a large number of hyperparameters and complex models like in the case of ensembles, it takes a lot of time to find the best set of parameters by hand. One such method used in case of hyperparameter tuning is Bayesian optimization which when applied to the models yields the best set of parameters.

11.2.5 Repeated K-Fold Cross-Validation

Evaluating a model is a core part of building an effective machine learning model. It provides a way to improve the estimated performance of a machine learning model. This implies simply running the multiple-time cross-validation process and recording the mean outcome from all runs through all folds. This mean outcome is assumed to be a more reliable calculation of the model's actual undiscovered mean output on the dataset, as measured using the standard error. The mean output recorded from a single run of cross-validation with k-fold could be noisy. The scikit-learn Python machine learning library provides an implementation of repeated k-fold cross-validation through the RepeatedKFold class. It provides the parameters, the number of folds (*n_splits*), which is the "*k*" in k-fold cross-validation, and the number of repeats (*n_repeats*) which can be specified according to the needs. So, the final values of different metrics were recorded using this technique. Figure 11.6 shows how k-fold cross-validation is performed.

Equation (11.9) shows how the final evaluation is done and generalized performance is calculated.

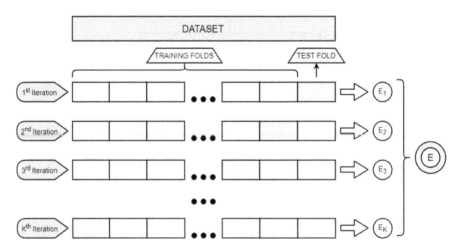

FIGURE 11.6
K-fold cross-validation.

$$\text{Performance} = \frac{1}{K} \sum_{a=1}^{K} P_a \qquad (11.9)$$

where,

K = number of folds

P = Any metric which is considered for measuring performance (example: mean squared error)

Performance = Final evaluation of the model, usually mean-squared error is selected as the parameter for judging performance.

Algorithm 1 explains the complete flow for the methodology proposed in this paper to predict the tumor size for melanoma.

ALGORITHM 1:   PIPELINE FOR PROPOSED METHOD

1: *do* {
2:   *DataPreprocessing()*
3:   *CreateModel()*
4:   *TrainModel()*
5:   *MakePredictions()*
6: }
7:
8: *function DataPreprocessing()* {
9:   *Data = Load_Dataset("/path/to/dataset")*
10:   *Data = Clean_Dataset(Data)*
11:   *FinalData = FeatureSelectionandEngineering(Data)*
12: }
13:
14: *function CreateModel()* {
15:   *List_Models = [XGBoost,LightGBM,ExtraForest,CatBoost,Bagging Regressor]*
16:   *FinalModel=create_model(list,parameters,mode="StackingRegressor")*
17:   *Return FinalModel*
18: }
19:
20: *function TrainModel(FinalModel)* {
21:   *CV = RepeatedKFold(n_splits = 10,n_repeats = 3)*
22:   *ScoringParameters = ['R2','MAE','MSE','MBE']*
23:   *Train,Validation = split_dataset(FinalData,val_size = 0.2)*
24:   *Performance = Cross_Validate(FinalModel,Train,Validation,CV, ScoringParameters)*

```
25:   FinalModel.finetune(Performance)
26:   FinalModel.save(MTSR)
27: }
28:
29: function MakePrediction() {
30:   model = load_model(MTSR)
31:   Data = load_data
32:   model.predict(Data)
33: }
```

11.3 Result

Different models were applied to the preprocessed dataset and the models were evaluated on different parameters. Bayesian optimization was used and the results were noted. Variations in results of different models were recorded and the model with the best fit was taken into account.

11.3.1 Bayesian Optimization

Bayesian optimization [30, 31] is a technique that uses Bayes' theorem to direct a search in order to find a minimum or maximum objective function. It is especially helpful when these tests are expensive, when one does not have access to derivatives, or when the problem at hand is not convex. Bayesian methods keep track of previous evaluation outcomes that they use to construct a probabilistic model mapping of hyperparameters with the probability of a score on the objective function $P(\text{score}/\text{hyperparameter})$.

This model is considered an objective function "surrogate" and is defined as $p(y/x)$. The surrogate is much easier to refine than the objective function and Bayesian methods operate by finding the next set of hyperparameters to measure the real objective function by choosing the hyperparameters that better perform the surrogate function. In other words, we need to construct a surrogate probability model of the objective function and then find the hyperparameters that better perform the surrogate after that add these hyperparameters to the true objective function. Then change the surrogate model that integrates new data. This process will be replicated until the maximum iteration is achieved. The goal of Bayesian reasoning is to become "less wrong" with more data than these methods do by continuously updating the surrogate probability model for each measurement of the objective function. Bayesian optimization approaches are effective when they allow rational decisions about the next hyperparameters. By testing hyperparameters that tend to be more promising from previous results, Bayesian approaches can find better model settings than random searches in less iterations. Bayesian model-based approaches can find better hyperparameters in less time because they justify the right set of hyperparameters to be tested on the basis of past trials.

Bayesian optimization performed on the different models produces the results as shown in the tables. Different range of set of parameters was fed into the function along with the model to be applied. It gave the suitable set of parameters that yield the best results on the target metric that was chosen and after which this can be applied for getting the best results on the test data for making the predictions. Results from Bayesian optimization are summarized in Tables 11.2, 11.3, and 11.4.

TABLE 11.2

Optimizing Bagging Tree Regressor

Iteration	target	max_features	min_samples	estimator
1	−3.953	0.2722	16.31	231.3
2	−3.906	0.806	19.94	181.8
3	−3.97	0.3485	20.44	387.4
4	−3.767	0.8875	10.23	250.3
5	−3.893	0.7144	18.39	211.1
6	−3.824	0.6045	13.57	104.1
7	−3.944	0.7948	22.3	209.5
8	−3.668	0.6532	3.734	210.6
9	−3.859	0.9389	16.98	219.2
10	−3.747	0.8091	9.287	270.4
11	−3.667	0.999	2.0	334.9
12	−3.668	0.999	2.0	142.0
13	−3.963	0.999	25.0	325.9
14	−3.968	0.999	25.0	135.0
15	−3.697	0.1493	2.014	399.6
16	−3.666	0.999	2.0	366.9
17	−3.67	0.999	2.0	301.2
18	−3.659	0.5486	2.078	258.1
19	−3.668	0.999	2.0	171.0
20	−3.678	0.999	2.0	100.0
21	−3.676	0.88	2.051	121.6
22	−3.668	0.999	2.0	227.3
23	−3.739	0.1363	2.138	382.2
24	−3.667	0.999	2.0	189.7
25	−3.655	0.7567	2.026	348.7

TABLE 11.3

Optimizing LightGBM Regressor

iteration	target	min_child_samples	estimators	num_leaves	subsample
1	−3.855	9.788	386.6	51.89	0.9141
2	−3.89	24.5	281.8	43.82	0.9207
3	−3.857	28.95	462.8	47.89	0.8004
4	−3.848	22.09	413.8	48.51	0.8245
5	−3.854	17.58	204.1	68.64	0.9531
6	−3.876	14.12	384.6	33.77	0.7475
7	−3.879	28.33	395.4	49.86	0.9155
8	−3.838	12.92	370.4	73.46	0.7745
9	−3.859	25.05	243.1	65.21	0.8818
10	−3.831	10.47	477.5	52.11	0.9637
11	−3.927	6.495	255.3	32.37	0.87
12	−3.895	19.87	360.0	32.17	0.8246
13	−3.897	13.24	350.9	35.59	0.8429
14	−3.863	19.15	202.0	60.87	0.9648
15	−3.826	24.76	497.6	77.94	0.9168
16	−3.843	12.13	387.5	53.9	0.6783
17	−3.878	14.56	216.2	52.58	0.9928
18	−3.84	8.099	235.8	66.93	0.8349
19	−3.885	16.79	232.1	41.46	0.96
20	−3.88	15.42	360.8	30.31	0.7203
21	−3.83	6.388	453.8	79.37	0.8022
22	−3.883	21.3	498.9	30.36	0.6823
23	−3.842	5.34	294.8	79.25	0.6585
24	−3.838	29.66	415.5	79.28	0.9079
25	−3.832	6.938	409.4	79.39	0.6229
26	−3.811	7.748	500.0	78.27	0.9518
27	−3.896	5.074	443.6	30.65	0.6585
28	−3.952	27.48	201.3	31.16	0.8138
29	−3.836	28.8	309.2	79.8	0.6058
30	−3.848	5.633	200.1	73.61	0.9991
31	−3.831	11.07	264.2	79.81	0.6659
32	−3.838	5.695	499.0	59.94	0.9259
33	−3.851	18.62	477.9	80.0	0.6438
34	−3.837	10.76	439.4	62.17	0.6376
35	−3.839	29.6	377.4	79.81	0.8218
36	−3.857	29.91	497.3	56.67	0.7261
37	−3.839	19.28	391.6	79.57	0.7044
38	−3.836	28.59	339.6	79.39	0.6017
39	−3.85	29.44	200.9	79.44	0.6483
40	−3.832	14.59	429.7	78.71	0.6268

TABLE 11.4

Optimizing XGBoost

Iteration	target	gamma	max_depth	min child_weight	estimators
1	−4.058	1.915	11.6	4.377	435.6
2	−4.044	2.726	8.488	8.019	487.4
3	−4.085	3.578	10.51	6.835	413.8
4	−4.074	5.612	10.53	0.1377	431.8
5	−4.077	3.649	11.54	0.7538	310.6
6	−4.095	6.514	9.575	7.887	295.1
7	−4.078	8.691	9.926	8.021	243.1
8	−4.05	7.046	7.969	9.249	332.6
9	−4.108	0.5981	7.659	0.4736	402.5
10	−4.143	5.333	6.39	5.614	298.9
11	−4.03	1.119	11.46	5.659	202.0
12	−4.051	9.121	13.11	9.921	487.6
13	−4.071	2.853	11.62	4.781	258.7
14	−4.068	0.5387	10.06	9.82	237.2
15	−4.082	7.385	11.29	4.716	232.1
16	−4.112	9.0	9.751	5.359	201.9
17	−4.054	4.369	11.51	9.182	387.7
18	−4.047	1.498	12.71	8.31	390.1
19	−4.042	1.526	11.12	5.282	485.4
20	−4.062	5.026	10.83	8.192	217.1
21	−4.073	9.958	14.5	1.311	355.5
22	−4.08	7.127	6.407	0.5419	499.4
23	−4.073	0.0	15.0	10.0	338.4
24	−4.058	0.0	15.0	10.0	500.0
25	−4.094	4.34	14.62	9.908	210.4
26	−4.04	9.308	14.12	9.868	323.6
27	−4.108	10.0	6.0	10.0	457.7
28	−4.136	9.816	14.74	0.207	276.9
29	−4.112	0.0	6.0	10.0	200.0
30	−4.077	0.3101	11.64	0.2621	210.2
31	−4.081	0.2288	13.88	3.747	202.6
32	−4.094	0.04106	6.1	4.463	363.0
33	−4.148	10.0	15.0	0.0	397.0
34	−4.183	0.0	15.0	0.0	458.7
35	−4.076	10.0	15.0	10.0	436.3
36	−4.108	10.0	6.0	10.0	270.1
37	−4.076	0.0	6.0	0.0	244.7
38	−4.024	0.0	15.0	10.0	281.2
39	−4.144	10.0	15.0	0.0	332.4
40	−4.11	10.0	6.0	10.0	356.9

11.3.2 Experimental Result and Comparison

The metrics used for evaluation were as follows:

Root mean square error (RMSE) is the standard deviation of the residuals (prediction errors). Residuals measure how far the data points are from the regression line. RMSE is a measure of how spread out these residuals are. In other words, it tells how concentrated the data is around the line of best fit. Root mean square error is commonly used in regression analysis to verify experimental results. The question used for calculation is as follows:

$$RMSE_{fo} = \left[\sum_{i=1}^{N} \sqrt{\frac{\left(z_{f_i} - z_{0_i} \right)^2}{N}} \right] \tag{11.10}$$

where,

$(z_{fi} - z_{oi})$ is the difference between the predicted and actual value, and N is the value of the sample size.

It can be seen that the XGBoost model has an RMSE value of 3.898, the LightGBM model gave the value as 3.804, and the Bagging Regressor gave 3.792. Using stacking regressor, XGBoost, LightGBM, Catboost, and Extra Tree Regressor were combined and it gave an RMSE value of 3.587 while XGBRFRegresssor, LightGBM, Extra Tree, and Bagging Regressor gave 3.572. But the best value was achieved by stacking up XGBoost, LightGBM, Catboost, Extra Tree, and Bagging Regressor as base models and taking Linear Regressor as the meta-model. The final value achieved was 3.564.

Mean absolute error (MAE) measures the average magnitude of the errors in a set of predictions, without taking their direction into account. It's the average over the test sample of the absolute differences between prediction and actual observation where all individual differences have equal weight. MAE is given by the equation:

$$MAE = \frac{\sum_{j=1}^{n} \left| y_j - \hat{y}_j \right|}{n} \tag{11.11}$$

where,

n is the number of samples.

While using k fold cross-validation and keeping MAE as the scoring parameter, it was found that Xgboost gave a 2.803 as the final value, whereas LightGBM gave a value of 2.642 and 2.788 in the case of Bagging Regressor. Using a stacking regressor, XGBoost, LightGBM, Catboost, and Extra Tree Regressor were combined and it gave an MAE value of 2.481 while XGBRFRegresssor, LightGBM, Extra Tree, and Bagging Regressor gave 2.473. But the best value was achieved by stacking up XGBoost, LightGBM,

Catboost, Extra Tree, and Bagging Regressor as base models and taking Linear Regressor as the meta-model. The final value achieved was 2.454 which is the least as compared to the other models.

R2 score is a statistical measure of how close the data points are to the fitted regression line. It is also known as the coefficient of determination, or the coefficient of multiple determination for multiple regression. R-squared values range between 0 and 1. R2 score of 0 indicates that the model explains none of the variability of the response data around its mean and 1 indicates that the model explains all the variability of the response data around its mean. So higher the value of the R2 score, the better the model will fit your data. XGBoost model got a value of 0.588 and on the other hand LightGBM got 0.608 value, whereas Bagging Regressor got a value of 0.611. A hybrid model of XGBoost, LightGBM, Catboost, and Extra Tree Regressor gave a value of 0.651, and stacking of XGBRFRegresssor, LightGBM, Extra Tree, and Bagging Regressor gave a value of 0.654. The best value was attained by the hybrid model of XGBoost, LightGBM, Catboost, Extra Tree, and Bagging Regressor that is 0.656.

Mean bias error (MBE) is typically meant to represent the overall "bias" model; that is, the average over-or under-prediction. MBE can provide valuable details but should be viewed carefully as it is inconsistently related to typical error magnitude, other than being an underestimate. Values of MBE for different models are given in Table 11.5. MBE is calculated using formula:

$$MBE = \frac{1}{N} \sum_{i=1}^{N} \hat{y}_i - y_i \qquad (11.12)$$

TABLE 11.5

Different Evaluation Metrics

S.No	Models	RMSE	MAE	R-Squared	MBE
1	XGBOOST	3.898	2.803	0.588	−0.0709
2	LightGBM	3.804	2.642	0.608	−0.0962
3	Bagging Regressor	3.792	2.788	0.611	−0.0973
4	XGBoost Cataboost LightGBM Extra Tree Regressor	3.587	2.481	0.651	0.0014
5	XGBRFRegressor LightGBM Extra Tree Regressor Bagging Regressor	3.572	2.473	0.654	−0.0187
6	XGBoost LightGBM Extra Tree Regressor Catboost Bagging Regressor	3.564	2.454	0.656	0.0005

where,
  $\hat{y}$ is the predicted value, and
  y is the observed value.

So, different models have been evaluated on the given dataset and the values of different metrics have been summarized in Table 11.5. It has been cleared from the results that the best fit was achieved by a hybrid model of XGBoost, LightGBM, Catboost, Extra Tree, and Bagging Regressor where these are base models and a Linear Regressor was taken as the meta-model.

11.4 Conclusion

Predicting the presence of melanoma tumor has been extensively covered in the previous existing literature; however, predicting the size of tumor has not been extensively covered. Therefore, in this chapter, we have proposed an approach for predicting the size of tumor with the help of given dataset that contains features such as the mass of the area understudy, the size of the area understudy, ratio of normal to the malign surface, unrecoverable area of skin damaged by the tumor, and others required for prediction of size. Various machine learning algorithms were tested on the given dataset but a stacking model of XGBoost, LightGBM, Extra Tree Regressor, Catboost, and Bagging Regressor provided the best results with 3.564 RMSE value, 2.454 MAE value, 0.656 R-squared value, and 0.0005 MBE value. This model outperformed all other approaches.

References

[1] Schadendorf D, van Akkooi AC, Berking C, Griewank KG, Gutzmer R, Hauschild A, et al. Melanoma. *The Lancet* 2018; 392(10151): 971e84.
[2] Gandini S, Sera F, Cattaruzza MS, Pasquini P, Zanetti R, et al. Meta-analysis of risk factors for cutaneous melanoma: III. Family history, actinic damage and phenotypic factors. *European Journal of Cancer* 2005a; 41: 2040–2059.
[3] Titus-Ernstoff L, Perry AE, Spencer SK, Gibson JJ, Cole BF, et al. Pigmentary characteristics and moles in relation to melanoma risk. *International Journal of Cancer* 2005; 116: 144–149.
[4] Esteva A, Kuprel B, Novoa RA, Ko J, Swetter SM, Blau HM, et al. Dermatologist-level classification of skin cancer with deep neural networks. *Nature* 2017; 542(7639): 115.

[5] Brinker TJ, Hekler A, Enk AH, Klode J, Hauschild A, Berking C, et al. Deep learning outperformed 136 of 157 dermatologists in a head-to-head dermoscopic melanoma image classification task. *European Journal of Cancer* 2019; 113: 47e54.

[6] Grzymala-Busse P, Grzymala-Busse JW, Hippe ZS. (2001, October). *Melanoma prediction using data mining system LERS*. In *25th Annual International Computer Software and Applications Conference. COMPSAC 2001* (pp. 615–620). IEEE.

[7] Sondermann W, Utikal JS, Enk A H, Schadendorf D, Klode J, Hauschild A, ... Haferkamp S. Prediction of melanoma evolution in melanocytic nevi via artificial intelligence: a call for prospective data. *European Journal of Cancer* 2019; 119: 30–34.

[8] Grzymala-Busse JW, Hippe ZS (2001). Melanoma prediction using k-Nearest Neighbor and LEM2 algorithms. In *Intelligent Information Systems 2001* (pp. 43–55). Physica, Heidelberg.

[9] Usher-Smith JA, Emery J, Kassianos AP, Walter FM. Risk prediction models for melanoma: a systematic review. *Cancer Epidemiology and Prevention Biomarkers* 2014; 23(8): 1450–1463.

[10] Milton MAA. Automated skin lesion classification using ensemble of deep neural networks in isic 2018: Skin lesion analysis towards melanoma detection challenge. 2019. *arXiv preprint arXiv:1901.10802.*

[11] Liu C, Zoph B, Neumann M, Shlens J, Hua W, Li LJ, Fei-Fei L, Yuille A, Huang J, Murphy K. (2018). *Progressive neural architecture search.* In *Proceedings of the European Conference on Computer Vision (ECCV)* (pp. 19–34).

[12] Szegedy C, Ioffe S, Vanhoucke V, Alemi A. (2016). Inceptionv4, inception-resnet and the impact of residual connections on learning.

[13] Hu J, Shen L, Sun G. (2018). *Squeeze-and-excitation networks.* In *Proceedings of the IEEE conference on computer vision and pattern recognition* (pp. 7132–7141).

[14] Autier P. Doré for Epimel and Eortc Melanoma Cooperative Group JF. Influence of sun exposures during childhood and during adulthood on melanoma risk. *International Journal of Cancer* 1998; 77(4): 533–537.

[15] Jiang MY, Dragnev NC, Wong SL. Evaluating the quality of reporting of melanoma prediction models. *Surgery.* 2020 May 21.

[16] Clark Jr WH, Elder DE, Guerry IV D, Braitman LE, Trock BJ, Schultz D, Synnestvedt M, Halpern AC. Model predicting survival in stage I melanoma based on tumor progression. *JNCI: Journal of the National Cancer Institute.* 1989; 81(24): 1893–1904.

[17] Vijayalakshmi MM. Melanoma skin cancer detection using image processing and machine learning. *International Journal of Trend in Scientific Research and Development (IJTSRD).* 2019; 3(4): 780–784.

[18] Mane SS, Shinde SV. Different techniques for skin cancer detection using dermoscopy images. *International Journal of Computer Sciences and Engineering* 2017; 5(12): 2347–2693.

[19] Li Y, Shen L. Skin lesion analysis towards melanoma detection using deep learning network, 2018. arxiv:1904.073653v2 [cs.cv]

[20] Razzak MI, Naz S and Zaib A. Deep learning for medical image processing: overview, challenges and future. 2018. arxiv:1852.3865v2 [cs.cv]

[21] Jain N, Jhunthra S, Garg H, Gupta V, Mohan S, Ahmadian A, Salahshour S, Ferrara M. Prediction modelling of COVID using machine learning methods from B-cell dataset. *Results in Physics* 2021; 21: 103813.

[22] Jain N, Gupta V, Shubham S, Madan A, Chaudhary A, Santosh KC. Understanding cartoon emotion using integrated deep neural network on large dataset. *Neural Computing and Applications*, 2021: 1–21.

[23] Chen T, Guestrin C. (2016, August). *Xgboost: A scalable tree boosting system*. In *Proceedings of the 22nd ACM SIGKDD international conference on knowledge discovery and data mining* (pp. 785–794).

[24] Friedman JH. Greedy function approximation: a gradient boosting machine. *Annals of Statistics* 2001: 1189–1232.

[25] Ridgeway G. Generalized boosted models: A guide to the gbm package. *Update*, 2007; 1(1), 2007.

[26] Ke G, Meng Q, Finley T, Wang T, Chen W, Ma W, … Liu TY. (2017). *Lightgbm: A highly efficient gradient boosting decision tree*. In *Advances in neural information processing systems* (pp. 3146–3154).

[27] Geurts P, Ernst D, Wehenkel L. Extremely randomized trees. *Machine Learning* 2006; 63(1): 3–42.

[28] Breiman L. Bagging predictors. *Machine Learning* 1996; 24(2): 123–140.

[29] Prokhorenkova L, Gusev G, Vorobev A, Dorogush AV, Gulin A. (2018). *CatBoost: unbiased boosting with categorical features*. In *Advances in Neural Information Processing Systems* (pp. 6638–6648).

[30] Snoek J, Larochelle H, Adams RP. (2012). *Practical bayesian optimization of machine learning algorithms*. In *Advances in Neural Information Processing Systems* (pp. 2951–2959).

[31] Pelikan M, Goldberg DE, Cantú-Paz E. (1999) *BOA: The Bayesian optimization algorithm*. In *Proceedings of the genetic and evolutionary computation conference GECCO-99* (Vol. 1, pp. 525–532).

12

A Fuzzy-Based Approach for Characterization and Identification of Sentiments

Madhav Kindra, Vikrant Dixit, and Vedika Gupta
Bharati Vidyapeeth's College of Engineering, New Delhi, India

CONTENTS

12.1 Introduction 219
12.2 Literature Survey 221
12.3 Overview of the Proposed Framework 224
12.4 Materials...225
    12.4.1 Linguistic Resources 225
    12.4.2 Dataset 227
12.5 Methodology 227
    12.5.1 Corpus Filtering and Preprocessing 227
    12.5.2 Word Study 228
    12.5.3 Sentiment Analysis Using Fuzzy Logic 229
        12.5.3.1 Developing Fuzzy Rules and Membership
                Function..230
        12.5.3.2 Establishing Rule Strength 230
        12.5.3.3 Aggregating Rule Strength and Output
                Membership Function...231
        12.5.3.4 Defuzzification 231
12.6 Results 232
12.7 Conclusion 233
References...234

12.1 Introduction

The importance of the society lies in the fact that it provides us with a system alongside a platform to work together for the betterment of the world as a whole. When society's efforts come together and become a collective force, we are able to improve our living and social conditions. One's surroundings and his vicinity also play a vital role along with the society to develop a healthy mindset for day-to-day activities. One has to take care of his mental, personal, emotional, and social well-being to stay healthy and continue to live his life to the fullest. Out of the above, emotional well-being is the one which drives the others to perfection. Ultimately, emotional aspects of any human being are crucial not only to himself but also to the others to comprehend. At this stage, we clearly see the importance of emotional intelligence, the tolerance, and capacity to be aware of, control, and express one's emotions in response to the stimuli, and to handle interpersonal relationships judiciously and empathetically. In order to master this art, one has to know himself and the environment in which he resides. And it is these emotions which drive the basic sentiments of users toward any perspective they face in the world. A sentiment can be considered as a belief which emerges from the emotions we go through.

A wide range of research work is related to sentiments, expressed by humans, in the fields of communication, psychology, linguistics, and uncountably more. Sentiments are categorized into two broad fields or polarity namely positive and negative. If any sentence contains no polarity, it is considered as a neutral sentence.

In the era of Internet and communications, sentiments are generally and greatly expressed in the form of text. People share their views and opinions digitally more than in person [1]. Blogs, news, reviews, articles, and much more are examples of such textual data. Text often reflects the writer's emotional state and this is a valuable resource for many purposes. The sentiments of the writer can be identified by using various methods. This analysis [2] is useful for a variety of purposes. Some domains where this technology [3] can be used include software engineering, website customization, education, gaming, etc.

In order to perceive and recognize sentiments from any text, any system [4] has to study the input text very carefully. It has wide applications in several fields majorly including:

1. Opinion Analysis: Opinion analysis aims to receive positive and negative comments about the product as reviews. It helps many companies to examine the opinion of their customers in a great manner.

2. Better Human-Computer Interaction Systems: In human-computer interaction systems, to recognize the sentiments of the human speech and to make the system feel like humans, sentiment analysis (SA) alongside emotion recognition techniques is applied. This has various uses in several industries and businesses' evolution.

3. Text to Speech Generation: To generate speech from a text, it is needed to analyze the sentiments expressed within the text. In this

way, the analysis of the sentiments from text becomes the constructive area of research for speech generation.

Our method deals with this situation and extracts sentiments from the textual data. Textual data is used for this operation and all the factors related to the sentiments are taken into consideration. The sentiments can be of positive or negative polarity. The sentences which are neutral in nature contribute a little or we can say that they do not change the polarity of the entire paragraph or the document and hence are not used in our analysis. Not only this, the fact that multiple sentiments can occur simultaneously is also taken into consideration. The process of SA is done using a layered framework which is discussed in the subsequent sections.

The rest of the chapter has been arranged in the following manner. Section 12.2 discussed the related work. Section 12.3 discusses the proposed framework and its working. Section 12.4.

12.2 Literature Survey

In recent years, SA is widely studied, and it is a subfield of natural language [5]. Capturing and understating sentiment that are behind the opinions is considerably important with the growth of social media platforms (blogs, Twitter, and social networks). SA is especially important which gathers huge amounts of data and applies some mythologies and techniques to extract opinions, sentiments, attitudes, and emotions from text/subject.

SA is used to detect the sentiment from given data with the help of natural language processing techniques [6]. A deeper and well-defined explanation of these fields is given in the below sections and described some recent related work.

In the past years, we have seen SA as one of the fastest growing areas which is using the natural language processing, text mining approach to extract the useful information from the word, sentence, and document to help in the decision-making process. SA can be defined as the process of studying the subjective knowledge in an expression, which can be opinions, mood, behavior, appraisals, emotions, attitudes, etc. toward a topic, person, or any other entities. Opinions or sentiment can be classified into three categories such as positive, negative, or neutral. For example: "I really like your personality" → Positive. Hence, we can say that SA is the method of classifying sentiment (positive, negative, or neutral) for entire documents.

SA, also called opinion mining [7] is a useful method which comes within natural language processing technique that allows us to identify the sentiment, and study subjective information. Both SA and opinion mining can be referred interchangeably because it comes under the same area of study. SA and opinion mining are almost the same thing. More specifically, SA identifies the sentiment in a word/expression in a text and then analyzes it whether opinion mining is used to extract and analyze people's opinion, mood, and feelings about an entity.

SA has been the most interesting area for researchers in recent years. We came up with an approach to achieve phrase-level SA [8]. Phrase-level SA is the process of identification of sentiments in terms of phrases, instead of the document-level SA. In phrase-level SA, we look for the sentiment of a phrase/entity; it also called as an entity-level SA. The need of studying phrase-level SA is to extract more different sentiments from a document, as has been outlined. Although identification of the local sentiment is more reliable and appropriate than the global document sentiment, an approach is used to identify semantic relationship between the sentiment and the subject, by determining the polarity of the sentiment in a subject.

SA processes mainly focus on the polarity [9] value of the given input (positive, negative, or neutral). SA uses various natural language processing techniques. Most commonly used method in SA includes: (1) rule-based approach which performs SA based on a given set of manually crafted rules and procedures, (2) automatic system processes that use machine learning techniques to learn from trained data, and (3) hybrid systems approach that combine both rule-based and automatic approaches and provide more appropriate solutions. Previously existing work mainly used three features: lexicon features, POS features, and micro-blogging features that are mostly used for SA and opinion mining-based study [10]. There are some other approaches which combine these three mentioned features, giving major emphasis to POS tags with or without word prior polarity involved. In SA, a positive document does not mean that the opinion holder is happy with all of its content (like if the document refers to several products); in the same way, a negative opinion, expressed for specific entities, does not conclude that the given data is fully negative (It could be positive and negative). There are various machine learning [11]-based classification algorithms that involve statistical models like naïve Bayes, logistic regression, support vector machines, commonly used in analysis of sentiment.

At this aim, SA works at the feature level: first, it will find nouns and noun phrases as features and apply lexicon-based method and fuzzy-logic method to determine the orientation of sentiment. SA focused on analyzing sentiment and recognizing the emotions or the opinions at the feature level [7, 12]. Detecting sentiment in text is quite a challenging task, yet there are various machine learning algorithms, deep learning and fuzzy logic based, which are used in identification and characterization of emotions. Some of them are shown in Table 12.1.

TABLE12.1

Various Researches Related to Emotion Recognition

Authors and Title	Technique	Approach	Dataset	Result/Performance
Indhuja and Raj Reghu (2014) [13]	Tree bank model	Fuzzy model	Product reviews (cars, movies, mobiles)	85.58% accuracy
Austermann, Anja, et al. (2005) [14]	Rule-based classification with fuzzy logic	ML	Real-time talking of people	84% accuracy
Liu & Cocea (2017, February) [15]	Rule-based classification	Fuzzy model	Movie reviews	95.1% accuracy
Adarsh (2019) [16]	Word clustering	ML	News headlines	Enhancement of current emotion recognition scores
Poria et al. (2016) [17]	Ensemble learning with CNN and linguistic approach	DL	Laptop and restaurant reviews	79.2% accuracy
Cambria, Erik, et al. (2011) [18]	Multidimensionality reduction	ML	Twitter, journal and patient reviews	89% accuracy
Lazemi, Soghra, and Hossein Ebrahimpour-Komleh et al. (2018) [19]	Multilabel learning methods	ML	IMDB reviews	Better performance than trivial models
Cambria, Erik, et al. (2012) [20]	Sentic computing	Linguistic approach with ML	Blogs	97% precision
Jain, Ubeeka, and Amandeep Sandhu. et al. (2009) [21]	Comparison study of various ML techniques	ML	Twitter dataset	Comparative study
Cambria, Erik, et al. (2011) [22]	Sentic computing	ML	blogs	73% precision
Subasic, Pero, and Alison Huettner. Et al. (2001) [23]	Semantic typing	Fuzzy model	Movie reviews and news headlines	Affects sets generated
Strapparava, Carlo, and Rada Mihalcea. et al. (2008) [24]	Knowledge and corpus-based annotation using Naive Bayes classifier	ML	News headlines and blogs	91.92% recall

12.3 Overview of the Proposed Framework

Figure 12.1 shows the logical schema of our proposed framework. The main purpose of this framework is to analyze the text for determining sentiment, including some additional participants such as WordNet, SentiWordNet, and VADER. There are several steps that need to be performed in order to get the desired output. First step is collecting a dataset; we need the input data which can be a web post, a document, a sentence, a word, or a paragraph which is taken from any textual data resource. When we collect the data, the next step is to clean that data which is called data cleaning. Data cleaning is the process where we dealt with removal of incorrect, corrupted, duplicate, or incomplete data within a dataset. Now further we have to perform data preprocessing which comes under corpus filtering which includes several methods like Lemmatization which is used to reduce the inflections (derived forms) of the word to, stop words removal that applies in order to perform filtering of unused words like is, the etc., POS Tagging used for the purpose of achieving label of the words in the sentence with the correct part of speech. After data preprocessing, synset association is to be performed in order to understand the proper meaning of words by applying lexicon such as VADER, WordNet, SentiWordNet, etc.

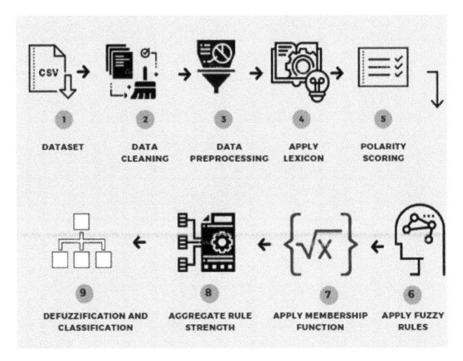

FIGURE 12.1
Block diagram of the framework employed.

Next phase of our model is word study, expression study. Each of these activities is parallelly used to achieve SA and emotion extraction. In the SA process, it includes identification and recognition of sentiment which is associated with the given words; measurement of polarity, assessment of positive, negative, and neutral score (in context of words). In word study, we have to calculate the polarity score. Polarity can be defined as emotions expressed in a sentence; polarity score means summing up the intensity value of each word present in the text. By calculating the polarity score of each word, we can easily determine the sentiment [25]. After that, we implemented expression study which comprises four main methods; these are apply fuzzy rules, membership function, determine aggregate rule strength, and then perform defuzzification and classification.

In analysis of sentiments, it involves finding patterns in the form of words which are related to sentiment and emotion extraction. These words include adverbs/adjectives that qualify the nouns and are also known as opinionated words. First step of expression study is to apply some fuzzy rules. This step deals with fuzzification which means decomposition of an input and/or output data into one or more fuzzy sets. After fuzzification, we have to calculate membership function; the triangular membership function for a fuzzy set S on the universe of discourse X is defined as $\mu_A : X \rightarrow [0,1]$, where each element of X is mapped to a value between 0 and 1. By applying some aggregate rule strength, we will get better results and after all that last step we have to perform is applying defuzzification and classification. Defuzzification is the process which is producing a quantifiable result with the help of Crisp logic, given fuzzy sets and corresponding membership degrees, whereas fuzzy classification can be defined as the process of combining individuals having the same characteristics into a fuzzy set.

Each phase of the proposed model tries to achieve main activities that are SA by performing various operations on different entities like words and sentences. Process starts with the processing of text first and then the result includes analysis of sentiment of the input text data; the calculated intensity value of each entity is expressed in the range$[0,1]$. Further sections will provide you with detailed explanations about each phase which is mentioned in Figure 12.1 for better and comprehensive understanding.

12.4 Materials

12.4.1 Linguistic Resources

This section defines the linguistic resources like WordNet, SentiWordNet, and VADER that we have used in our proposed model:

(i) **WordNet:** WordNet [26] is a large collection of lexical databases for the English language. The purpose behind using WordNet is used to link words into semantic relations like synonyms, hyponyms, and meronyms. In WordNet, we associate every single word with one or more synsets (a set of one or more synonyms that are interchangeable). In other words, we can say that WordNet is a group of words which is linked by lexical and semantic relations. WordNet includes nouns, verbs, adjectives, and adverbs which are organized into a set of synonyms. Each synset in WordNet has a separate and unique identified index and provides the meaning and a definition. WordNet is a combination and extension of a dictionary and thesaurus. WordNet labels the semantic relations among words. WordNet has the most common frequent relation between synsets which is called the super-subordinate relation (also known as hyponymy or ISA relation). WordNet is widely used in many applications; examples are question answering, machine translation and information retrieval automatic text classification, and many more.

(ii) **SentiWordNet**: SentiWordNet [27] is the most common resource for lexical study used for opinion mining-based and SA-based applications. SentiWordNet is a lexical database which is commonly used in opinion mining. It is derived from the WordNet database where each word is associated with numerical scores such as positive and negative sentiment information. In SentiWordNet, we assign each synset of WordNet with three sentiment numerical scores: positivity, negativity, objectivity. SentiWordNet positive and negative values lie in the range [0,1], and their sum is 1.0 for each synset.

SentiWordNet has been created automatically from the combination of linguistic and statistical classifiers. It is basically the result of an annotation of synsets of wordnet based on the notation of positivity, negativity, and neutrality. In SentiWordNet, each term of the synset has the same opinion-related property and different senses of a term may have different opinion-related properties. Although SentiWordNet used to perform opinion mining and SA mainly computes opinion polarity at the sentence level, SentiWordNet provides SA-based information which is extracted from the given words and then matched to calculate an overall score and perform prediction operation of the desired expression.

(iii) **VADER**: VADER [28] (Valence Aware Dictionary and Sentiment Reasoner) is a lexicon resource that can be defined as a rule-based SA tool which is used to attune to sentiments expressed in given data. Vader became very popular in recent years because it not only gives you the positivity and negativity score of a particular word but also tells you about how positive or negative a sentiment is. VADER calculates sentiment score of text by summing up the intensity value of each word present in the text.

For example, words like *"upset," "sad," "unhappy." "dislike"* all these words convey a negative sentiment. The VADER SA tool is smart enough to understand the basic context of these words, such as *"did not upset"* as a positive sentiment statement. Vader has several advantages: It is working excellently on social media type text and generates desired accurate results.

12.4.2 Dataset

SA is a technique which can be used on a variety of datasets. These include textual data, audios, video data, conversational data between humans or even humans and HCI, and a lot more. In our approach to determine the sentiment, we have made use of the largest and widest available source of data on the Internet, which is undoubtedly textual data.

We have used a pre-labeled dataset [29] which has sentences labeled as positive and negative based on what sentiment they have. The sentiment score is either 1 (for positive sentiment score) or 0 (for negative sentiment score). The sentences have been taken from IMDB's website having only positive and negative polarity sentences. For this dataset, there exists 500 positive and 500 negative sentences for our analysis. Moreover, all of these sentences have been selected randomly from a much wider dataset that was procured earlier.

An attempt was also made to select only the positive and the negative sentiment score sentences in order to avoid the neutral sentiment scores. This helped a lot during the analysis of the sentiment score for our proposed method.

12.5 Methodology

12.5.1 Corpus Filtering and Preprocessing

The first phase of the text analysis is divided into four steps:

1. Lemmatization
2. Stop words removal
3. POS tagging
4. Synset association

In the first step, we performed Lemmatization to reduce the inflections (derived forms) of the word to their root words or the Lemmas. The root word produced in this step is a dictionary form word with meaning. Then in the stop word removal process, all the words that do not contain any specific meaning to them, i.e., stop words, are filtered out from the sentence. Redundant words like "is, am the" are removed as they are not needed in the processing of the text. POS Tagging is executed to label words in a sentence with the correct part of speech. In order to associate proper POS tags to every word of the text, we used the help of Stanford POS tagger [30] to analyze every word in the dataset.

Synset association is the most crucial and complicated part of this phase. This section forms the base for further analysis in our model. Human languages are highly ambiguous due to words having several interpretations based on their usage. Before associating a synset to a word, it is important to determine the correct "sense" of the word with respect to its usage in the sentence. To solve this semantic ambiguity, it is necessary to design a means to identify the circumstances in which a word is used. In other words, we can say this section is aimed at performing Word Sense Disambiguation.

Our methodology makes use of the Python Implementations of Word Sense Disambiguation (pywsd) package for WordNet association [31]. The package implements established related measures to gain knowledge about the synsets. For proper synset assignment, we calculate the similarity values of each word present in the sentence. To compute similarity values, we take two words at a time. We measure the relatedness between all the synsets of the two words. The synset of a word that gives the maximum similarity value with the other words present in the sentence will be assigned to the word. To find the similarity values, we have utilized Wu and Palmer Similarity [32].

However, WordNet can calculate similarity values only between nouns. To find similarity values of other parts of speech present in the sentence (adjective, verbs, etc.) we must "substantivize" the word. A substantive word plays the role of a noun in a sentence. The aim is to find a semantic category in WordNet that encapsulates the context in which the word is used in the sentence.

12.5.2 Word Study

The next phase after text processing is Word Study. It involves the analysis of the text on word level. For sentiment extraction, the polarity value of each word needs to be determined. For this purpose, we have compared two lexical resources – SentiWordNet and VADER.

SentiWordNet is an easy-to-use resource which is readily available via the NLTK platform. It maps a word with its corresponding WordNet synsets and assigns a polarity value to them. The polarity score has three categories: positive score, negative score, and objective score. Therefore, each synset will have a positive score, negative score, and objective score associated with it. A synset is considered positive if the associated positive score is greater than the negative score, and negative if vice-versa is true.

The setback we faced while using SentiWordNet was that, at times, it was producing very obscure results. Many of the synsets in SentiWordNet do not have any positive or negative score attached to them. As a result, it becomes hard to derive any meaningful insight into the sentiment of such words.

VADER is an open-source sentiment lexicon that was created in 2014. It is a rule-based, sentence-level SA tool that gives excellent results on social media data. The heuristics for Vader are based on the properties and characteristics of the text that affect the intensity of the sentiment. VADER has five identifiers that indicate a change in the intensity of the sentiment. These identifiers are as follows: punctuations, capitalization, degree modifiers, presence of conjunction, and negation [28]. The VADER heuristic has the treatment for these intensifiers. VADER also has fast and economic computation while providing high accuracy. Considering all the above factors, we chose to use VADER in our implementation.

After analyzing the text, VADER investigates if the words in the text are present in its lexicon. The polarity score for each word has three different values: positive sentiment, negative sentiment, and neutral sentiment. The scores range from +4 to −4.

12.5.3 Sentiment Analysis Using Fuzzy Logic

Finally, the analysis of sentiments of the proposed model deals with finding out patterns in the form of words that contribute to sentiment and emotion extraction. These words include adverbs/adjectives that qualify the nouns and are also known as opinionated words. One such way to identify patterns is POS tagging of sentences.

One of the ways to calculate polarity is by summing the polarity of individual words belonging to the pattern. But this fails in cases where some special adverbs such as very, extremely, really, always, simply, etc. modify the intensity of the polarity of words that follow them like very good, extremely bad, etc. Other adverbs like never, not, etc. change the polarity of the following adjective like the cases not good will be bad. Finally, there are combinations of them such as not very good, and not too bad.

Considering human language that is plainly spoken and is highly ambiguous, it therefore becomes more crucial to identify the language semantics. In such scenarios, fuzzy techniques help to manipulate the uncertainty and vagueness enclosed in the language expressions that otherwise could be avoided or ignored in the more traditional dichotomic modeling.

The Fuzzy Inference System (FIS) used here is the most popular Mamdani Fuzzy Inference System [33]. It was proposed in 1975 by Ebhasim Mamdani and works on following principles:

1. Developing Fuzzy rules and membership function.
2. Establishing rule strength.
3. Aggregating rule strength and output membership function.
4. Defuzzification.

12.5.3.1 Developing Fuzzy Rules and Membership Function

This step deals with fuzzification of positive and negative scores for each review that we obtained using VADER [28]. The triangular membership function for a fuzzy set S on the universe of discourse X is defined as μ_A: $X \rightarrow [0,1]$, where each element of X is mapped to a value between 0 and 1. It is defined by a lower limit a, an upper limit b, and a value m, where $a < m < b$.

The graphical representation of triangular membership function is shown in Figure 12.2. Or more compactly,

$$\mu_A(x) = max\Big(min\big((x-a)/(b-a),(c-x)/(c-b)\big),0\Big) \qquad (12.1)$$

For computing positive, negative, and output variables, three fuzzy sets are defined as low: [min,min,mid], medium: [min,mid,max], and high: [mid,max,max] using triangular membership functions. The range for three fuzzy sets on the scale for negative, neutral, and positive is as follows:

Output negative is *opNeg* = [0,0,5]
Output neutral is *opNeu* = [0,5,10]
Output positive is *opPos* = [5,10,10]

12.5.3.2 Establishing Rule Strength

The following nine rules are formulated in a way that the common scores results in neutral sentiment otherwise consider the highest score for positive or negative as sentiment.

$$\mu_A(x) = \begin{cases} 0, & x \le a \\ \dfrac{x-a}{m-a}, & a < x \le m \\ \dfrac{b-x}{b-m}, & m < x < b \\ 0, & x \ge b \end{cases}$$

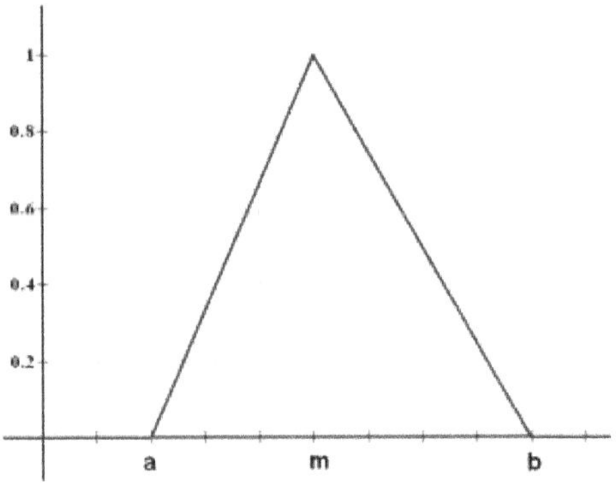

FIGURE 12.2
Fuzzy triangular membership function.

Fuzzy rules are defined as follows:

		Negative Score		
		LOW	MEDIUM	HIGH
	LOW	Rule1 Neutral	Rule4 Negative	Rule7 Negative
Positive Score	MEDIUM	Rule2 Positive	Rule5 Neutral	Rule8 Negative
	HIGH	Rule3 Positive	Rule6 Positive	Rule9 Neutral

12.5.3.3 Aggregating Rule Strength and Output Membership Function

For negative value

$$neg1 = max\left(Rule8, max\left(Rule4, Rule7\right)\right)$$

$$outputLow = min\left(neg1, opNeg\right) \tag{1}$$

For neutral value

$$neu1 = max\left(Rule9, max\left(Rule1, Rule5\right)\right)$$

$$outputMid = min\left(neu1, opNeu\right) \tag{2}$$

For positive values

$$pos1 = max\left(Rule6, max\left(Rule2, Rule3\right)\right)$$

$$outputHigh = min\left(pos1, opPos\right) \tag{3}$$

The resultant membership function (for negative, neutral, and positive, respectively), i.e., output_low, output_mid, and output_high are computed by clipping the membership functions of consecutive parts with overall rule strength as shown in Equation (1), (2), and (3).

12.5.3.4 Defuzzification

The method used for defuzzification is center of gravity/centroid of area method [34]. It provides crisp values based on COG. The total area of membership function is divided into subareas. The summation of all the COGs for each subarea is calculated to find the crisp value. The defuzzified output is classified into following categories:

$$Negative : 0 < \left(outputLow\right) < 3.33$$

$$Neutral: 3.34 < (outputMid) < 6.66$$

$$Positive: 6.67 < (outputHigh) < 10$$

12.6 Results

The evaluation of our performance results can be done on the basis of the polarity that was detected from the sentences. Figure 12.3 presents how a single review is processed by our fuzzy model on IMDB dataset having 500 positive and negative sentences.

Figure 12.4 is the visualization of triangular membership function described in the universe of discourse: positive is (0-1), negative is (0-1), and output (0-10). Here blue color is for negative, green color for neutral, and red color for positive. The aggregated output has been calculated by using the centroid of area method. The defuzzified output shown in bold is equal to 4.91 making the sentiment score of sentence neutral. For the measurement of performance, we took into account the precision score of our model. Precision score after applying fuzzy to all the sentences of the dataset came out to be 80.25%.

---

club leader reacted accordingly---------------------------------{'neg': 0.0, 'neu': 1.0, 'pos': 0.0, 'compound': 0.0}

Positive Score for each tweet :
0.0

Negative Score for each tweet :
0.0

Firing Strength of Negative (wneg): 0.0
Firing Strength of Neutral (wneu): 1.0
Firing Strength of Positive (wpos): 0.0

Resultant consequents MFs:
op_activation_low: [0. 0. 0. 0. 0. 0. 0. 0. 0. 0.]
op_activation_med: [0. 0.2 0.4 0.6 0.8 1. 0.8 0.6 0.4 0.2]
op_activation_high: [0. 0. 0. 0. 0. 0. 0. 0. 0. 0.]

Aggregated Output: [0. 0.2 0.4 0.6 0.8 1. 0.8 0.6 0.4 0.2]
Defuzzified Output: 4.91

Output after Defuzzification: Neutral

FIGURE 12.3
Sample review using VADER lexicon and fuzzy logic.

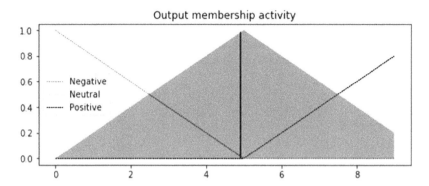

FIGURE 12.4
Output showing sentiments.

12.7 Conclusion

In the recent years, the growth of SA is one of the most active research areas for the researchers due to different reasons such as it helps in better decision making, improves understanding of human emotions, etc. SA is getting more attention from many researchers in the past ten years because of many reasons. First, it is widely used in many applications, in almost every domain; second, huge amount of data is recorded every day in digital form from various Web sources. Emotion is very crucial part in our day-to-day life because it governs all the human actions and emotions: in decision-making processes; understanding the human emotions from different Web sources; a judgment about an entity, an advertisement, a movie, etc. which can be used in a better way to improve business plans, market behavior, advertisement, etc. It can also be helpful in conveying the appropriate opinion.

To provide better solutions, we have proposed this model framework that aims to provide a better way to do SA, which detects opinion from text, in terms of sentiment polarity. In this chapter, we have used a fuzzy-based model to detect the emotions. The use of fuzzy logic allows providing more accurate results. Fuzzy modifiers are an adequate tool to tune these emotions in the text. We have compared our proposed framework against the various existing systems and observed that this approach framework attains higher performance for all the cases.

In the future, we would like to explore the other dimensions of our proposed model framework and improve the analysis and synthesis of the extracted sentiment in terms of other entities like audio, video, etc. We have provided a solution for SA; in future, we will expand this work to analyze emotion. In this proposed model, we have used fuzzy-based model; we will redefine and improve this model by using some additional features to compute the polarity of emotions in different entities like pattern, sentence, etc.; we would like to build an intelligent application which is capable of processing the text, interpreting it, and expressing the emotions associated with the human interaction or involvement.

References

[1] K. Kannan, M. Goyal, & G. Jacob. (2012). Modeling the impact of review dynamics on utility value of a product, *Social Network Analysis and Mining* 3(3), 401–418.

[2] E. Cambria, A. Hussain, C. Havasi, & C. Eckl (2010). Sentic computing: exploitation of common sense for the development of emotion-sensitive systems. In A. Esposito, N. Campbell, C. Vogel, A. Hussain, & A. Nijholt (Eds.), *Development of Multimodal Interfaces: Active Listening and Synchrony, Lecture Notes in Computer Science* (Vol. 5967, pp. 148–156). Springer, Berlin, Heidelberg.

[3] R.W. Picard. (1997). *Affective Computing*. MIT Press, Cambridge, MA.

[4] E. Cambria, R. Speer, C. Havasi, & Hussain, A. (2010). *Senticnet: A Publicly Available Semantic Resource for Opinion Mining*.

[5] Cambria, E., Schuller, B., Xia, Y., & Havasi, C. (2013). New avenues in opinion mining and sentiment analysis, *IEEE Intelligent Systems* 28(2), 15–21.

[6] Gupta, V., Jain, N., Katariya, P., Kumar, A., Mohan, S., Ahmadian, A., & Ferrara, M. (2021). An emotion care model using multimodal textual analysis on COVID-19. *Chaos, Solitons & Fractals*, 144, 110708.

[7] Liu, B. (2012). Sentiment analysis and opinion mining. *Synthesis Lectures on Human Language Technologies*, 5(1), 1–167.

[8] Wilson, Theresa, Janyce Wiebe, and Paul Hoffmann. (2005). *Recognizing contextual polarity in phrase-level sentiment analysis. Proceedings of Human Language Technology Conference and Conference on Empirical Methods in Natural Language Processing*.

[9] Bravo-Marquez, F., Mendoza, M., & Poblete, B. (2013). *Combining strengths, emotions and polarities for boosting Twitter sentiment analysis*. In *Proceedings of the Second International Workshop on Issues of Sentiment Discovery and Opinion Mining*.

[10] H. Saif, He, Y., & Alani, H. (2012). *Semantic sentiment analysis of twitter*. In *Proceedings of the 11th International Conference on The Semantic Web, ISWC'12*, vol. PartI, Springer-Verlag, Berlin, Heidelberg, pp. 508–524.

[11] Mudinas, A., Zhang, D., & Levene, M. (2012). *Combining lexicon and learning based approaches for concept-level sentiment analysis*. In *Proceedings of the first international workshop on issues of sentiment discovery and opinion mining*.

[12] M. Cataldi, A. Ballatore, I. Tiddi, M.-A. Aufaure. (2013). Good location, terrible food: detecting feature sentiment in user-generated reviews. *Social Network Analysis and Mining*.

[13] Indhuja, K., & Reghu, R.P.C. (2014). *Fuzzy logic based sentiment analysis of product review documents*. In *2014 First International Conference on Computational Systems and Communications (ICCSC)*. IEEE.

[14] Austermann, A., et al. (2005). *Fuzzy emotion recognition in natural speech dialogue. ROMAN 2005*. In *IEEE International Workshop on Robot and Human Interactive Communication, 2005*. IEEE.

[15] Liu, H., & Cocea, M. (2017, February). *Fuzzy rule based systems for interpretable sentiment analysis*. In *2017 Ninth International Conference on Advanced Computational Intelligence (ICACI)* (pp. 129–136). IEEE.

[16] Adarsh S. R. (2019). Enhancement of text based emotion recognition performances using word clusters. *International Journal of Research – Granthaalayah*, 7(1), 238–250.

[17] Poria, S., Cambria, E., & Gelbukh, A. (2016). Aspect extraction for opinion mining with a deep convolutional neural network. *Knowledge-Based Systems*, 108, 42–49.

[18] Cambria, E., et al. (2011). *Isanette: A common and common sense knowledge base for opinion mining*. In *2011 IEEE 11th International Conference on Data Mining Workshops*. IEEE.

[19] Lazemi, S., & Ebrahimpour-Komleh, H. (2018). *Multi-emotion extraction from text based on linguistic analysis*. In *2018 4th International Conference on Web Research (ICWR)*. IEEE, 2018.

[20] Cambria, Erik, et al. (2012) Sentic computing for social media marketing. *Multimedia Tools and Applications* 59(2), 557–577.

[21] Jain, U., & Sandhu, A. (2009). A review on the emotion detection from text using machine learning techniques. *Emotion*, 5(4), 2645–2650.

[22] Cambria, E., et al. (2011). *Sentic medoids: Organizing affective common sense knowledge in a multidimensional vector space*. In *International Symposium on Neural Networks*. Springer, Berlin, Heidelberg.

[23] Subasic, P., & Huettner, A. (2001). Affect analysis of text using fuzzy semantic typing. *IEEE Transactions on Fuzzy Systems* 9(4), 483–496.

[24] Strapparava, C., & Mihalcea, R. (2008) *Learning to identify emotions in text*. In *Proceedings of the 2008 ACM symposium on Applied Computing*. ACM.

[25] Piryani, R., Gupta, V., & Singh, V. K. (2018). Generating aspect-based extractive opinion summary: Drawing inferences from social media texts. *Computación y Sistemas*, 22(1), 83–91.

[26] Miller, G. A. (1995). Wordnet: a lexical database for english, *Communications of the ACM*, 38(11), 39–41.

[27] Esuli, A., & Sebastiani, F. (2006, May). *Sentiwordnet: A publicly available lexical resource for opinion mining*. In *LREC* (Vol. 6, pp. 417–422).

[28] Hutto, C. J., & Gilbert, E. (2014). *Vader: A parsimonious rule-based model for sentiment analysis of social media text*. In *Eighth international AAAI conference on weblogs and social media*.

[29] Kotzias et al. (2015). *From Group to Individual Labels using Deep Features*. In *KDD '15: Proceedings of the 21th ACM SIGKDD International Conference on Knowledge Discovery and Data Mining KDD*.

[30] Toutanova, K., Klein, D., Manning, C.D., & Singer, Y. (2003, May). *Feature-rich part-of-speech tagging with a cyclic dependency network*. In *Proceedings of the 2003 conference of the North American chapter of the association for computational linguistics on human language technology-volume 1* (pp. 173–180). Association for Computational Linguistics.

[31] Tan, L. (2014) Pywsd: Python implementations of word sense disambiguation (wsd) technologies [software]. Retrieved from https://github.com/alvations/pywsd.

[32] Wu, Z., & Palmer, M. (1994, June). *Verbs semantics and lexical selection.* In *Proceedings of the 32nd annual meeting on Association for Computational Linguistics* (pp. 133–138). Association for Computational Linguistics.

[33] Srivastava, R., & Bhatia, M. P. S. (2013, August). *Quantifying modified opinion strength: A fuzzy inference system for sentiment analysis.* In *2013 International Conference on Advances in Computing, Communications and Informatics (ICACCI)* (pp. 1512–1519). IEEE.

[34] Wang, W. J., & Luoh, L. (2000). Simple computation for the defuzzifications of center of sum and center of gravity. *Journal of Intelligent & Fuzzy Systems, 9*(1,2), 53–59.

13

Fingerprint Alterations Type Detection and Gender Recognition Using Convolutional Neural Networks and Transfer Learning

Gaurav Kataria, Akansh Gupta, V. Sirish Kaushik, Gopal Chaudhary, and Vedika Gupta

Bharati Vidyapeeth's College of Engineering, New Delhi, India

CONTENTS

13.1 Introduction 237
13.2 Related Work 239
13.3 Contributions.. 241
13.4 Materials and Methods 242
    13.4.1 Dataset 242
    13.4.2 Methods 245
        13.4.2.1 AlexNet.. 245
        13.4.2.2 AkaNet 245
        13.4.2.3 VGG-16 246
13.5 Results 247
    13.5.1 Classification Accuracies 247
    13.5.2 Alteration Detection.. 247
    13.5.3 Alteration Type Detection 247
    13.5.4 Gender Recognition 248
13.6 Conclusion 252
References.. 253

13.1 Introduction

Biometrics are classified into physiological biometrics and behavioural biometrics. Physiological biometrics include traits such as fingerprint, face, iris, and voice. These days, biometrics are generally preferred over the traditional token-based systems, such as identity cards, keys or knowledge-based systems, such

DOI: 10.1201/9781003134138-13

as passwords [1]. As an individual may forget their passwords, consequently they may reuse the same passwords, which overall affects the reliability of the system in a negative way. Fingerprint, a physiological biometric, is widely used these days in smartphones for login, financial transaction authentication, Aadhar cards and is even used by Law Enforcement and Border Control agencies for identification and authentication of individuals [2].

In the context of investigation of a crime scene, fingerprints are present in abundance, and also evidence of fingerprints being used to identify criminals and tracking their illicit activities has been found by archaeologists in the Qin Dynasty (221–206 BC). Hence, fingerprints are extremely important when it comes to forensic science. Fingerprints are unique patterns made by friction ridges and furrows where friction ridges include loops, whorls and arches. These features are distinct for everyone and are the basis for fingerprint recognition. But fingerprints can be altered and sometimes are intentionally altered by criminals to evade Law Enforcement and Border Control agencies. Fake fingers are particularly used by criminals to adopt someone else's identity, whereas fingerprint alteration is typically done to mask one's own identity. Here alteration of fingerprint means the damaging of friction ridge patterns. Alteration types are classified on the basis of the flow of dermatoglyphic crests into obliteration and distortion [3]. Here, distortion can be further classified into z-cut and central rotation as seen in Figure 13.1.

Obliterations are the most common type of alteration. Obliterations are carried out by abrasion, application of chemicals, burns or cuts. Distortion means a sudden change of orientation of the ridges along the scar produced by the alteration. This is done by removing a small section of skin from the fingertips and then replanting them in a different position. Distortion can be classified into central rotation or z-cut on the basis of the method used for replanting fingerprints. In central rotation, the patch, even from other fingers, is re-planted in place of rotation. In z-cut, a Z-shaped cut on the fingertip is made. Two triangular skin patches are lifted and switched and stitched back together [4]. The most common alteration type based on examination of ridge patterns are obliteration and distortion which makes up 89% and

FIGURE 13.1
Fingerprints with obliteration, central rotation and z-cut alterations, respectively.

10%, respectively, with 1% of alterations being imitations [2]. Apart from alteration type detection, gender recognition is also important as it helps in short listing of suspects hence reducing time for matching of fingerprints and overall providing real-time identification of individuals [5]. An individual of the name Alexander Guzman was arrested in the US state of Florida in 1995 for having a fake passport [3]. Alexander had altered his fingerprints by making a z-cut. FBI after 2 weeks of investigation was able to reconstruct the characteristics of the fingerprints and eventually discovered the actual identity of the individual was of Jos Izquirdo who was a drug dealer. Hence, it is necessary to develop an efficient fingerprint recognition system that can identify altered images and alteration type and other fingerprint characteristics consequently, countering such illegal actions of criminals [6]. When it comes to 2D image recognition and analysis, deep learning is the most successful technique being highly accurate and fast, as it reduces the complexity of images and testing time hence providing precise and real-time fingerprint classification. Hence, the purpose of this research is to develop a deep learning approach which can detect alteration and type of alteration of fingerprints and moreover recognise the gender of the individual accurately and in real time. Authors of this paper propose to achieve this objective by employing CNN and transfer learning architectures to classify fingerprint images from the Sokoto Coventry Fingerprint dataset.

In this paper, we present deep learning models that can identify altered images, type of alteration present and gender, with high classification accuracy on the Sokoto Coventry fingerprint dataset. This research paper is organised into seven sections: Section 13.1 includes a general overview of fingerprints, types of fingerprint alteration and the importance of designing a reliable and efficient fingerprint recognition system. Section 13.2 explores other significant research work done in the area of fingerprint recognition, fingerprint alteration's type detection and gender classification. Section 13.3 presents the objective of the paper. Section 13.4 elucidates the methodology of the paper, mentioning the dataset used and explaining the deep learning models employed in this paper. Section 13.5 presents the results achieved by each model and Section 13.6 concludes the research paper. References are listed in the last section.

13.2 Related Work

The problem of identification of individuals on the basis of altered fingerprints has been explored by many researchers. Here are some research works related to our paper. Shehu et al. [2] proposed convolutional neural networks to identify the type of alteration. Their approach achieved a classification accuracy of 98.55%. They had split the dataset into 50% training and 50% testing and rescaled the images to 200×200 using bipolar interpolation. Their proposed approach consisted of five convolutional layers employing ReLU activation function. The first three convolutional layers were followed by max-pooling layers and the CNN model also consisted of two fully connected layers. They also employed a pre-trained residual CNN model on ImageNet which achieved an accuracy of 99.86%. Giudice et al. [3] employed Inception v3 architecture for detection of altered fingerprints, identification of type of alterations and recognition of gender, hand and fingers. Their proposed approach achieved an accuracy of 98.21%, 98.46%, 92.52%, 97.53% and 92.18% on the classification of fakeness, alterations, gender, hand and fingers, respectively, on the Sokoto Coventry Fingerprint dataset. Iloanusi et al. [5] proposed the use of CNNs for gender classification from fingerprints. The authors of the paper proposed a fusion scheme which significantly improved the classification accuracies when compared to their default approach. Their proposed approach achieved validation accuracies of 94.7%, 88.0% and 91.3% on male, female and overall classification, respectively. They utilised two datasets with one of them being the Sokoto Coventry Fingerprint dataset. Through their analysis, they concluded that validation accuracy of gender classification depends on the finger type.

Zhao et al. [7] proposed deep learning approaches to recover defective license plates and fingerprint images. They employed a denoising convolutional autoencoder to address defocus blurs in license plates and fingerprint images. They artificially generated images by generating artificial blurs and smudges that resemble real-world degradations. They performed Artificial Linear Smear Correction and Artificial Gaussian Smear Correction to generate clear images. They achieved a validation loss of 0.185 and 0.041 on linear smear model and Gaussian smear model, respectively. Narayanan et al. [8] proposed a time-domain approach which was used to find out the gender of a particular fingerprint obtained using systematic pixel counting. They achieved a classification accuracy of 90.2% and 96.4% for females and males, respectively, on the Sokoto Coventry Fingerprint dataset. Their proposed approach includes pre-processing images, extracting images, threshold calculation and comparison based on threshold. Rim et al. [9] employed AlexNet, VGG-16, Yolo-v2 and ResNet-50 for left-right hand classification, finger classification, sweat pore classification and scratch classification. Their Yolo-v2 model provided highest classification accuracy of 90.98%, 78.68% and 66.55% on left-right hand classification, scratch classification and finger classification, respectively, whereas their ResNet-50 model achieved the highest accuracy of 91.29% on sweat pore classification. Moreover, these models were efficient, taking 250.37 per image. Rim et al. created their own dataset which consists of 1069 fingerprints where 1008 and 61 belonged to Cambodian and Korean people, respectively. Authors of this paper aim to improve finger classification accuracy by collecting more fingerprints and aim to classify fingerprints into four categories by a single trained weight. Shehu et al. [10] proposed a CNN model which achieved an accuracy of 75.2%, 76.72% and 93.5% on gender, fingers and hand classification, respectively. They aim to further increase the classification accuracy of gender and finger recognition by using other publicly available datasets and introducing the use of minutiae points in the dense layers.

Yoon et al. [21] argued that NFIQ and similar software are not able to detect altered fingerprints efficiently. Hence, they proposed a technique for automatic detection of altered fingerprints based on examining the orientation field and minutiae distribution of fingerprints. They employed SVM classifier and utilised the NIST SD4 dataset containing 4,433 altered fingerprints. They stated that their proposed approach achieved a classification accuracy of 66.4% on alteration detection, whereas NFIQ algorithms only achieved an accuracy of 26.5% on the same. Tabassi et al. [23] proposed the use of CNN architectures such as Inception-v3 and MobileNet-v1 for alteration detection and GANs to produce synthetic altered fingerprints to increase the image dataset for training purposes. They utilised a dataset containing 4815 altered fingerprints and 4815 unaltered fingerprints from 270 subjects to train their CNN models. Their proposed CNN models achieved a true detection rate of 99.24% at a FDR of 2%. The authors of the paper made the synthetically generated altered fingerprint images open sourced as well. Patil et al. [26] utilised multimodal biometric characteristics including fingerprints, face and iris for gender recognition. They utilised the SDUMLA-HMT dataset. Their proposed approach included feature level fusion. They employed LDA, KNN and SVM classifiers. MB-LBP and BSIF feature descriptors have been fused together to extract features for multi-biometric traits. Their proposed approach of fusion of MB-LBP and BSIF features and employing SVM classifier achieved a classification accuracy of 99.8% on gender classification. The research gap of some of the research work mentioned above is that some of the datasets used consist of synthetically altered fingerprint images as there is a lack of publicly available datasets with real fingerprint alterations. Hence, a number of researchers have explored and experimented on this topic, utilising popular datasets like SOCOFING and NIST SD4, employing deep learning architectures, using fusion schemes and examining as well as extracting relevant features from fingerprints to improve overall accuracy.

13.3 Contributions

Biometrics are used widely for identification and authentication of individuals by government agencies and service providers. Law Enforcement agencies have been using fingerprints to identify and track criminals for centuries.

With the advent of technology in modern times, alterations in fingerprints is a possibility that can help criminals evade such agencies. The objectives of the paper are as follows.

1. Develop an efficient convolutional neural network model that can classify fingerprints as real or altered.
2. Identify nature or type of alterations made, if any.
3. Determine the gender of the individual concerned, thereby helping the agencies in countering such illegal actions of criminals.

13.4 Materials and Methods

The project involved multiple steps, starting with the dataset being imported from Kaggle. As a part of pre-processing, normalisation and undersampling were performed. Thereafter, the dataset was divided into training and testing sets which were in the ratio of 75% and 25%. We have employed six deep learning models. Each model was trained for 15 epochs, with training and testing batch sizes of 32. After training and testing, the validation accuracy of models AlexNet, MobileNet, VGG16, Inception-v3, ResNet-18 and AkaNet are calculated. The following subsections further explain the above stages in detail. Figure 13.2 below shows the various steps involved in our approach.

13.4.1 Dataset

We are using Sokoto Coventry Fingerprint dataset [11] which is a publicly available dataset [12] which was created with the sole aim of enhancing the security of biometric fingerprints such that criminals on the watchlist can be identified and apprehended even if their fingerprints are altered. This dataset consists of 6000 fingerprints belonging to 600 African subjects. There are ten fingerprints per subject and all subjects are 18 years or older. There are three types of altered fingerprints present in the dataset: obliteration, z-cut and central rotation. These alterations are synthetically altered versions of the fingerprints and are generated by the STRANGE toolbox.

There are a total of 55,273 images in the dataset where the resolution of each image is 96×103. We will be employing CNN and transfer learning models to classify fingerprints as real or altered, identify the type of alteration and also determine the gender of the individual. Further details about the dataset are mentioned in Table 13.1. Figures 13.3 and 13.4 show the fingerprint images from the above-mentioned dataset [2].

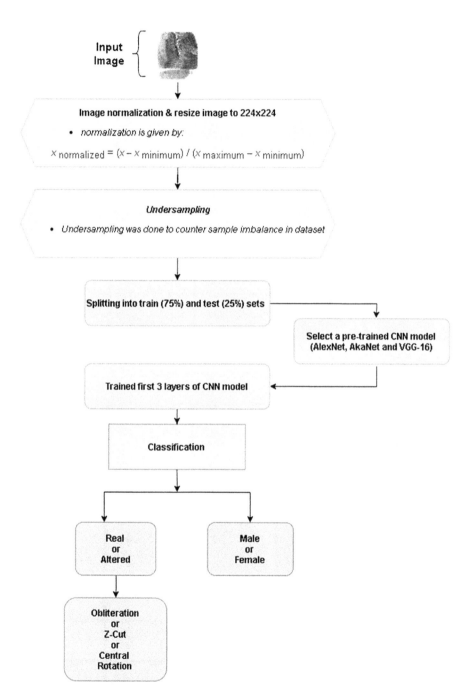

FIGURE 13.2
Flowchart of proposed methodology.

TABLE 13.1

Details about Sokoto Coventry Fingerprint Dataset

Data Format	BMP
Total no. of instances	55,273
No. of subjects	600
Instances per subject	10
Image size	96×103
Size of dataset	838 MB
Total unaltered images	6000
Levels of alteration	Easy, medium, hard
Alteration types	Z-cut, obliteration, central rotation

FIGURE 13.3
Images of five left hand fingerprints belonging to the same subject [2].

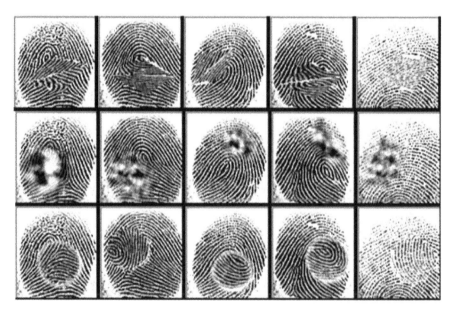

FIGURE 13.4
Images above, after being altered into z-cut, obliteration and central rotation, respectively [2].

13.4.2 Methods

As convolutional neural networks retain spatial information through filter kernels [13], in this work, we exploit this unique ability of CNNs to train multiple models to classify images from the Sokoto Coventry Fingerprint dataset into real or altered, central rotation, obliteration or z-cut and male or female.

13.4.2.1 AlexNet

AlexNet is a CNN model designed by Alex Krizhevsky, which achieved a top five error of 15.3% on the ImageNet dataset, while competing in the ILSVRC 2012 [9]. AlexNet introduced ReLU activation function, overlapping pooling layers and multi-gpu training, making it a highly efficient deep learning model. The model consists of five convolutional layers employing ReLU activation function, where some were followed by max-pooling layers. The model consists of three dense layers employing ReLU activation function and 0.5 dropout to overcome overfitting. We employed AlexNet and trained it from scratch on the Sokoto Coventry fingerprint dataset. The architecture of AlexNet is given below in Figure 13.5.

13.4.2.2 AkaNet

The convolutional neural network model presented in this paper, AkaNet as shown in Figure 13.6, consists of two convolution layers with filter size of 3 and employing ReLU activation function. Each of the convolutional layers is followed by a max-pooling layer with stride of 2. Followed by these layers are four dense layers with the first three employing ReLU activation function. Dropout was also employed to reduce overfitting. In AkaNet, Adam optimiser was employed and the learning rate of the model was 0.01. Learning rate is given by Equation 13.1:

$$LR = \lambda / 1 + (\omega \times \theta) \tag{13.1}$$

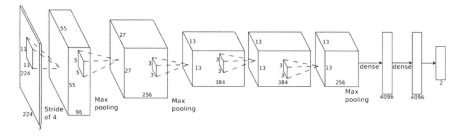

FIGURE 13.5
AlexNet's architecture.

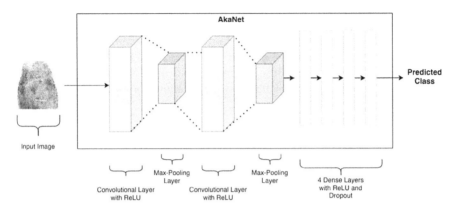

FIGURE 13.6
AkaNet's architecture.

Here λ is the initial learning rate, ω is the decay factor and θ is the current epoch. Loss function employed is cross entropy which is given by Equation 13.2:

$$E = -\sum_{c=1}^{M} y_{o,c} \log\left(p_{o,c}\right) \qquad (13.2)$$

13.4.2.3 VGG-16

VGG-16 is a deep learning model which was presented at Image-Net Large-Scale Visual Recognition Challenge 2014 by K. Simonyan and A. Zisserman [14]. VGG-16 attained a top 5% test classification accuracy of 92.7% on the ImageNet dataset which contains 14 million images belonging to 1000 classes. VGG-16 was the first model to introduce the use of multiple 3×3 kernel-sized filters one after the other replacing large kernel-sized filters. This feature helped the model to recognise and understand more complex features and patterns. The learning rate of the model was 0.0005, Adam optimiser was utilised and cross entropy loss function was used. The architecture of VGG-16 is given below in Figure 13.7.

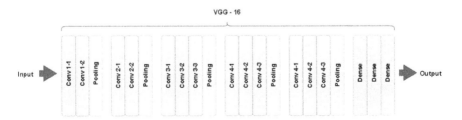

FIGURE 13.7
VGG16's architecture.

13.5 Results

CNN and transfer learning models such as AlexNet, AkaNet and VGG16 were trained and tested on the Sokoto Coventry Fingerprint dataset. These models classified fingerprints into real or altered, alteration types and gender of the individual. Results achieved and comparative analysis are mentioned in the following sub-headings.

13.5.1 Classification Accuracies

Accuracy attained by our proposed models is given below in Table 13.2:

13.5.2 Alteration Detection

Alteration detection is a binary classification problem with the objective of classifying between real or altered fingerprints. Giudice et al. [3] achieved a classification accuracy of 98.21%. As given in the Table 13.2, AlexNet achieved the highest classification accuracy of 98.50%. Hence, AlexNet outperformed the model proposed by Giudice et al by a small margin. AkaNet also achieved a high accuracy of 95.85%, deeming both of these models as accurate and efficient deep learning models usable for real-world applications. Model loss graph and confusion matrix of AlexNet are given below in Figure 13.8 and Table 13.3, respectively. Figure 13.9 below shows the ROC curve graph of AlexNet.

13.5.3 Alteration Type Detection

Alteration type detection is a ternary classification problem with the objective of classifying an altered fingerprint into obliteration, z-cut and central rotation. Giudice et al. [3] achieved a classification accuracy of 98.46% on alteration type detection, and Shehu et al. [2] achieved a classification accuracy of 98.55%. Models presented in this paper achieved the following classification

TABLE 13.2

Presents Classification Accuracies of Alteration Detection, Alteration Type Detection and Gender Classification Models

Model	Classification Accuracy		
	Alteration Detection	Alteration Type Detection	Gender Classification
AlexNet	98.50%	94.84%	83.07%
VGG16	78.52%	64.50%	55.60%
AkaNet	95.85%	73.80%	75.24%

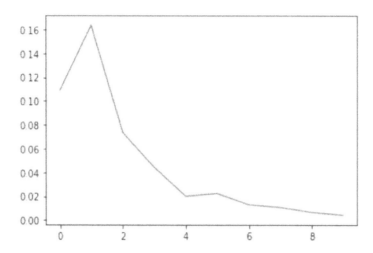

FIGURE 13.8
Model loss graph for AlexNet on alteration detection.

TABLE 13.3

The Confusion Matrix of AlexNet on Alteration Detection

	Predicted Altered	Predicted Real
True Altered	1491	9
True Real	37	1476

accuracies. From the above table, we can see that again AlexNet achieved the highest classification accuracy of 94.84%. But the accuracy achieved is lower than the accuracies achieved by Shehu et al. and Giudice et al. Model loss graph and confusion matrix of AlexNet on alteration type detection are given below in Figure 13.10 and Table 13.4, respectively. Figure 13.11 displays the ROC curve graph of AlexNet on alteration type detection.

13.5.4 Gender Recognition

Gender classification is a binary classification problem with the purpose of discriminating between the gender of the individual. Shehu et al. [10] achieved a classification accuracy of 75.2%, whereas Iloanusi et al. [5]

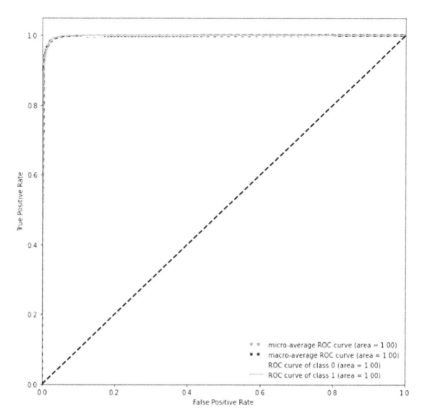

FIGURE 13.9
ROC curve graph of AlexNet on alteration detection.

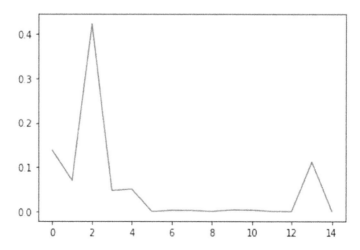

FIGURE 13.10
Model loss graph for AlexNet on alteration type detection.

TABLE 13.4

The Confusion Matrix of AlexNet on Alteration Type Classification

	Predicted CR	Predicted Obl	Predicted Real	Predicted Z-Cut
True CR	648	0	1	1
True Obl	3	645	0	2
True Real	12	2	606	30
True Z-Cut	34	0	53	563

Note: In the Table, CR is central rotation, Obl is obliteration.

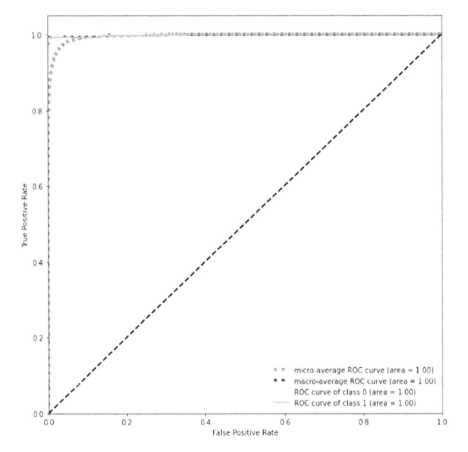

FIGURE 13.11
ROC curve graph of AlexNet on alteration type detection.

achieved a classification accuracy on gender recognition of 83.3% (without fusion) and 91.3% (with fusion). Proposed approach of Giudice et al. [3] achieved an accuracy of 92.52%. Accuracy achieved on gender classification by our proposed models is given below. AlexNet achieved an accuracy of 83.07%, hence outperforming Shehu et al. [10]. Model proposed by Iloanusi et al. [5] without fusion and our AlexNet model attained similar accuracies. Because of the sample imbalance present in the dataset, the classification accuracy has been affected. Model loss graph and confusion matrix of AlexNet on gender recognition are given below in Figure 13.12 and Table 13.5, respectively. Figure 13.13 shows the ROC curve graph of AlexNet on gender recognition.

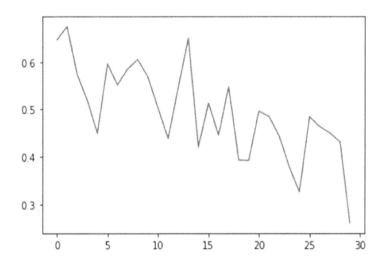

FIGURE 13.12
Model loss graph for AlexNet on gender recognition.

TABLE 13.5

The Confusion Matrix of AlexNet on Gender Recognition

	Predicted Female	Predicted Male
True Female	444	73
True Male	108	409

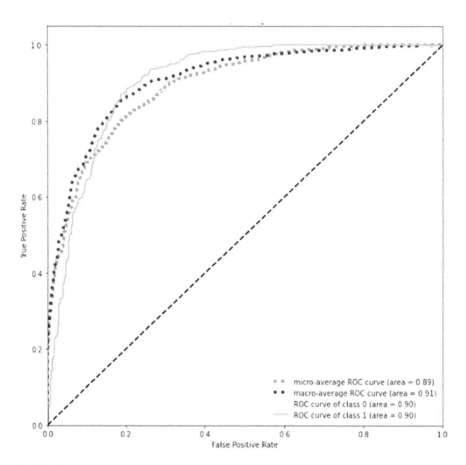

FIGURE 13.13
ROC curve graph of AlexNet on gender classification.

13.6 Conclusion

Alteration of fingerprints by criminals poses a huge security risk when it comes to authentication of an individual's identity. Apart from this, identification and analysing fingerprint characteristics such as gender of the individual can be helpful in criminal investigations. In this paper, the Sokoto Coventry fingerprint dataset was used, CNN and transfer learning models were employed to classify fingerprints as real or altered, identify nature or type of alterations and also the gender of the individual. AlexNet achieved the highest accuracies of 98.50%, 94.84% and 83.07% on alteration detection, alteration types detection and gender recognition, respectively. This approach outperformed the approach presented by Giudice et al. on alteration detection and also outperformed the approach presented by Shehu et al. on gender recognition. Hence, we can use such deep learning models for digital automatic detection of altered fingerprints and identification of gender.

However, one of the limitations of this work is that the proposed CNN and transfer learning models were evaluated on synthetically altered images due to the lack of publicly available datasets containing actual altered images. In future work, we would like to include other fingerprint datasets as well as explore other deep learning approaches to further improve the classification accuracy.

References

[1] Minaee, S., Abdolrashidi, A., Su, H., Bennamoun, M., & Zhang, D. (2019). Biometric recognition using deep learning: A survey. *arXiv preprint arXiv:1912. 00271.*

[2] Shehu, Y. I., Ruiz-Garcia, A., Palade, V., & James, A. (2018, October). Detection of fingerprint alterations using deep convolutional neural networks. In Kůrková V., Manolopoulos Y., Hammer B., Iliadis L., Maglogiannis I. (eds) *Artificial Neural Networks and Machine Learning – ICANN 2018.* ICANN 2018. Lecture Notes in Computer Science, vol. 11139. Springer, Cham. https://doi. org/10.1007/978-3-030-01418-6_6.

[3] Giudice, O., Litrico, M., & Battiato, S. (2020). *Single architecture and multiple task deep neural network for altered fingerprint analysis. 2020 IEEE International Conference on Image Processing (ICIP)*, 2020, pp. 813–817, doi:10.1109/ICIP40778. 2020.9191094

[4] Cummins, H. (1935) Attempts to alter and obliterate finger-prints. *Journal of American Institute of Criminal Law and Criminology*, 25(6), 982–991. doi:10.2307/1134845

[5] Iloanusi, Ogechukwu N., & Ejiogu, Ugochi C. (2020). Gender classification from fused multi-fingerprint types. *Information Security Journal: A Global Perspective*, 29(5), 209–219.

[6] Gottschlich, C., Mikaelyan, A., Olsen, M. A., Bigun, J., & Busch, C. (2015, September). Improving fingerprint alteration detection. In *2015 9th International Symposium on Image and Signal Processing and Analysis (ISPA)* (pp. 83–86). IEEE. doi:10.1109/ISPA.2015.7306037

[7] Zhao, Y., & Lebak, S. (2019). Deep Convolutional Autoencoder for Recovering Defocused License Plates and Smudged Fingerprints.

[8] Narayanan, A., & Sajith, K. (2019). *Gender Detection and Classification from Fingerprints Using Pixel Count (August 28, 2019).* In *Proceedings of the International Conference on Systems, Energy & Environment (ICSEE) 2019*, GCE Kannur, Kerala.

[9] Rim, B., Kim, J., & Hong, M. (2020). Fingerprint classification using deep learning approach. *Multimedia Tools and Applications*, 1–17.

[10] Shehu, Y. I., Ruiz-Garcia, A., Palade, V., & James, A. (2018β, December). *Detailed identification of fingerprints using convolutional neural networks.* In *2018 17th IEEE*

International Conference on Machine Learning and Applications (ICMLA) (pp. 1161–1165). IEEE.

[11] Shehu, Y. I., Ruiz-Garcia, A., Palade, V., & James, A. (2018). Sokoto coventry fingerprint dataset. *arXiv preprint arXiv:1807.10609.*

[12] https://www.kaggle.com/ruizgara/socofing

[13] Jain, R., Nagrath, P., Kataria, G., Kaushik, V. S., & Hemanth, D. J. (2020). Pneumonia detection in chest X-ray images using convolutional neural networks and transfer learning. *Measurement, 165,* 108046.

[14] Simonyan, K., & Zisserman, A. (2014). Very deep convolutional networks for large-scale image recognition. *arXiv preprint arXiv:1409.1556.*

[15] Szegedy, C., Vanhoucke, V., Ioffe, S., Shlens, J., & Wojna, Z. (2016). *Rethinking the inception architecture for computer vision.* In *Proceedings of the IEEE Conference on Computer Vision and Pattern Recognition* (pp. 2818–2826).

[16] He, K., Zhang, X., Ren, S., & Sun, J. (2016). *Deep residual learning for image recognition.* In *Proceedings of the IEEE Conference on Computer Vision and Pattern Recognition* (pp. 770–778).

[17] Howard, A. G., Zhu, M., Chen, B., Kalenichenko, D., Wang, W., Weyand, T., … Adam, H. (2017). Mobilenets: Efficient convolutional neural networks for mobile vision applications. *arXiv preprint arXiv:1704.04861.*

[18] Pettersson, M., & Obrink, M. (2001). Ensuring integrity with fingerprint verification. *Precise Biometrics White Paper, Lund, Sweden.*

[19] Papi, S., Ferrara, M., Maltoni, D., & Anthonioz, A. (2016, September). *On the generation of synthetic fingerprint alterations.* In *2016 International Conference of the Biometrics Special Interest Group (BIOSIG)* (pp. 1–6). IEEE.

[20] Feng, J., Jain, A. K., & Ross, A. (2010, August). *Detecting altered fingerprints.* In *2010 20th International Conference on Pattern Recognition* (pp. 1622–1625). IEEE.

[21] Yoon, S., Feng, J., & Jain, A. K. (2012). Altered fingerprints: Analysis and detection. *IEEE Transactions on Pattern Analysis and Machine Intelligence, 34*(3), 451–464.

[22] Kim, J., Rim, B., Sung, N. J., & Hong, M. (2020). A Comparative Study on the Effective Deep Learning for Fingerprint Recognition with Scar and Wrinkle. *Journal of Internet Computing and Services, 21*(4), 17–23.

[23] Tabassi, E., Chugh, T., Deb, D., & Jain, A. K. (2018, October). *Altered fingerprints: Detection and localization.* In *2018 IEEE 9th International Conference on Biometrics Theory, Applications and Systems (BTAS)* (pp. 1–9). IEEE.

[24] Uliyan, D. M., Sadeghi, S., & Jalab, H. A. (2020). Anti-spoofing method for fingerprint recognition using patch based deep learning machine. *Engineering Science and Technology, an International Journal, 23*(2), 264–273.

[25] Yuan, C., Xia, Z., Jiang, L., Cao, Y., Wu, Q. J., & Sun, X. (2019). Fingerprint liveness detection using an improved CNN with image scale equalization. *IEEE Access, 7,* 26953–26966.

[26] Patil, A., Kruthi, R., & Gornale, S. (2019). Analysis of multi-modal biometrics system for gender classification using face, iris and fingerprint images. *International Journal of Image, Graphics and Signal Processing, 11*(5), 34–43.

[27] Costa, H. S., Bellon, O. R., Silva, L., & Bowden, A. K. (2016, September). *Towards biometric identification using 3D epidermal and dermal fingerprints.* In *2016 IEEE International Conference on Image Processing (ICIP)* (pp. 3937–3941). IEEE.

[28] Kumar, N. (2019). Cancelable Biometrics: a comprehensive survey. *Artificial Intelligence Review, 53,* 3403–3446.

[29] Yang, W., Wang, S., Hu, J., Zheng, G., & Valli, C. (2019). Security and accuracy of fingerprint-based biometrics: A review. *Symmetry, 11*(2), 141.

[30] Gomez-Barrero, M., & Galbally, J. (2020). Reversing the irreversible: A survey on inverse biometrics. *Computers & Security, 90*, Article No 101700.

[31] Zhang, Y., Huang, Y., Wang, L., & Yu, S. (2019). A comprehensive study on gait biometrics using a joint CNN-based method. *Pattern Recognition, 93*, 228c236.

[32] Ross, A., Banerjee, S., Chen, C., Chowdhury, A., Mirjalili, V., Sharma, R., ... Yadav, S. (2019, June). *Some research problems in biometrics: The future beckons.* In *2019 International Conference on Biometrics (ICB)* (pp. 1–8). IEEE.

[33] Pfeuffer, K., Geiger, M. J., Prange, S., Mecke, L., Buschek, D., & Alt, F. (2019, May). *Behavioural biometrics in VR: Identifying people from body motion and relations in virtual reality.* In *Proceedings of the 2019 CHI Conference on Human Factors in Computing Systems*, New York, NY (pp. 1–12).

14

Content-Based Image Retrieval Using Intelligent Techniques

Prashant Srivastava

NIIT University, Rajasthan, India

Manish Khare

Dhirubhai Ambani Institute of Information and Communication Technology, Gandhinagar, Gujarat, India

Ashish Khare

University of Allahabad, Prayagraj, Uttar Pradesh, India

CONTENTS

14.1 Introduction 257
14.2 State of the Art...259
14.3 CBIR Using Intelligence Techniques 261
    14.3.1 The Proposed Method 263
    14.3.2 Experiment and Results 264
    14.3.3 Retrieval Result..265
    14.3.4 Performance Comparison 266
14.4 Conclusion and Future Scope 268
References...268

14.1 Introduction

The present era is witnessing proliferation of information at an exponential rate. Today, the exchange of information in various forms is taking place at a rapid pace. Out of the various forms of information, multimedia content has become the most popular among the people. Multimedia, which involves image, audio, and video content, is being shared on a massive scale. With the invention of various social media platforms, there has been a tremendous increase in this. Due to the presence of huge amount of multimedia content,

DOI: 10.1201/9781003134138-14

the need for efficient multimedia information retrieval (MIR) system has become very important. Users around the world are exchanging multimedia data extensively on a daily basis. Out of the various multimedia content, image is one such type of information, which is hugely popular among users of social media. Since the time the images were first used to share information, they have been a great interest of research to the scientists across the world. Various image processing techniques have been proposed to extract useful information from the image. Presence of social media platforms and other Web sources have led to the production and exchange of large number of images. For easy and efficient access of images, it is imperative to have a system which can automatically arrange and organize the images. Image retrieval systems overcome the challenges of automatic arrangement so as to make the accessing of images easy.

Image retrieval refers to searching and retrieval of images based on the features of image. Image retrieval system can be classified into two broad categories such as text-based image retrieval (TBIR) system and content-based image retrieval (CBIR) system. TBIR systems search for images based on keywords and text associated with the image. The retrieval accuracy depends on how users express the query. The expression of query by a user may differ from another user, as a particular user may use certain keywords according to his/her interpretation. Hence, TBIR systems may not always produce relevant results. Another limitation of text-based retrieval system is that they require manual annotation of large number tagging of enormous number of images, which is a tedious task.

Other than text-based retrieval system, CBIR system is another type of image retrieval system. CBIR is the process of searching and retrieval of image based on features of image such as color, texture, and shape. In a CBIR system, either the sketch of image or the image itself is provided as query. CBIR systems perform feature extraction from the query image thereby constructing a feature vector. The feature vector of query image is matched with the feature vector of images stored in the database in order to retrieve visually similar images. Such systems require no manual annotation of large numbers of images and are capable of retrieving visually similar images. The features of image play an important role in CBIR. The features used for retrieval may be either low-level features such as color, texture, and shape or high-level features such as semantic features.

The term CBIR first came into existence in the late 1980s [1]. Since then, a number of CBIR techniques have been proposed based on low-level features. Early CBIR systems were mostly based on exploitation of single features only. However, as the images became more complex due to invention of high-resolution camera, single features started proving to be insufficient to retrieve multiple features from image. This limitation was overcome by combination of low-level features which produced better retrieval results than single feature [2–4].

The combination of features proved to be quite efficient in extracting multiple features from images. However, these features have been mostly exploited on single resolution of image. There are varying level of details present in an image, and in order to extract them, single resolution of image is not considered to be sufficient. To overcome this limitation, the concept of multiresolution processing was introduced. The usage of multiresolution techniques shifted the trend of CBIR techniques from single-resolution processing-based techniques to multiresolution processing-based techniques. The advantage of using multiresolution processing of image is that the features which are left undetected at one scale get detected at another scale. A number of CBIR techniques based on multiresolution technique have been proposed which exploit feature description at multiple resolutions of image to construct feature vector [5–8].

The techniques discussed above mostly exploit low-level features at either single resolution or multiple resolutions of image. These techniques rely on visual features such as color, texture, shape, and spatial features to construct feature vector and utilize it to retrieve visually similar images. However, human beings use high-level or semantic features to recognize an image instead of low-level features. This results in a huge gap between how machines recognize an image and how we human beings interpret the same image. This gap is known as semantic gap [9]. Since human beings are capable of recognizing an image semantically due to their intelligence, there is a need for making machines intelligent so as to make them interpret and analyze any image just like human beings do. In other words, instead of using low-level features to construct feature vector, there is a need to use intelligent techniques which construct feature vectors using high-level or semantic features in order to recognize an image semantically.

The rest of the chapter is organized as follows: Section 14.2 discusses state-of-the-art in the field of CBIR. Section 14.3 discusses CBIR using intelligent techniques and finally Section 14.4 concludes the chapter.

14.2 State of the Art

The process of CBIR involves feature extraction followed by construction of feature vector. The features of image being used to construct feature vector are considered to be of primary importance. Early CBIR techniques mostly relied on extraction of primary features such as color, texture, and shape to construct feature vector. CBIR techniques based on color feature generally use color histogram [10, 11], color correlogram [12], and color coherence vector [13] to construct feature vector. Colors are visible features of an image and are invariant to certain geometric transformations. Apart from color feature, texture is another feature which has been extensively used for feature extraction. Texture feature has been mostly extracted using local pattern-based features. A number of local patterns have been proposed which efficiently extract texture features in an image [14–17]. Local patterns efficiently extract structural arrangement of intensity values by encoding relationship among intensity values in local neighborhood. Apart from color and texture, shape is another feature which has proved to be an efficient low-level feature. Shape descriptors such as moments [18, 19], histogram of oriented gradient (HOG) [20, 21], and polygonal structures [22], have been extensively used to extract shape feature from image.

The low-level features prove to be efficient in extracting visual features in an image. However, the structure of image is complex as it consists of multiple objects of different types. Therefore, low-level features used as single feature proved to be inefficient to extract multiple features in an image. This drawback was overcome by combination of primary features. The combinations of color and texture [23, 24], colour and shape [25], texture and shape [26], and colour, texture, and shape [27] have been proposed to extract multiple features in an image. The combination of features performs better than single feature as they combine advantages of multiple features and overcome limitations of each other.

The structure of image is complex consisting of varying level of details. It consists of foreground as well as background objects and large as well as small objects. The technique discussed above performs feature extraction from single-resolution processing of image. Single-resolution processing of image fails to extract varying level of details. This lacuna is overcome by multiresolution processing of images. Multiresolution processing of images refers to analysis and interpretation of images at more than one resolution. Multiresolution techniques tend to decompose images into multiple scales. The feature extraction is performed at different levels of resolution. The advantage of decomposing an image into multiple resolutions is that the features that are left undetected at one scale get detected at another scale. The decomposition of image into multiple resolutions is done using multiple resolutions such as wavelet transform, curvelet transform, and contourlet transform. Out of these multiresolution techniques, wavelet transform has been one of the most extensively used techniques for constructing feature vector for CBIR. Wavelet transform has been used either as single feature or in combination with primary features [28–37] for constructing feature vector for CBIR. Wavelets compute coefficients in multiple orientations and at multiple scales. These properties of wavelets help in construction of efficient feature vectors which contain features extracted at multiple resolutions and in multiple directions. However, wavelets suffer from certain drawbacks. Wavelets have limited directionality and are nonanisotropic in nature. Multiresolution techniques such as curvelet transform and contourlet transform overcome these drawbacks. These techniques compute coefficients at multiple orientations and are highly anisotropic in nature. Due to these properties, they construct more efficient feature vectors than wavelet transform and have been extensively used in the field of CBIR [38–45]. Multiresolution techniques are considered to be better than single-resolution techniques not only in terms of feature extraction but also in producing high retrieval accuracy.

14.3 CBIR Using Intelligence Techniques

CBIR techniques can be broadly classified into two categories: low-level feature-based techniques and high-level or semantic feature-based techniques. The techniques based on low-level features capture visual characteristics of image such as color, texture, and shape efficiently. However, such techniques may not always yield accurate results. Such techniques fail to recognize an image based on types of objects, events, scene, or situation being depicted in an image [46]. Human beings have tendency to use semantic features rather than primitive features to recognize an image. This results in a huge semantic gap between how human beings recognize an image and how machines interpret the same image. There are certain categories of image objects which appear in different varieties of color, texture, and shape such as chair and table. There is a possibility that a new version of such objects may occur in future [47]. Recognition of images containing such objects is difficult solely based on primitive features. This requires intelligence-based technique for extracting semantic contents from images. Human beings are capable of recognizing an image based on semantic contents as they are intelligent. This justifies usage of intelligent techniques for bridging semantic gap given the fact that the human brain is considered to be an intelligent object which has the capability to learn from the changes taking place in the environment. The technique used for identifying such image must provide a facility to incorporate new features when a newer version of object appears. Hence, for searching and retrieval of such types of images, an external knowledge base is required which not only stores wide variety of features of any object but also provides the facility to include newly encountered features when a newer version of same objects present in the image appears. The advantages of using intelligent techniques for CBIR are as follows:

1. Intelligent techniques help in bridging the semantic gap. There exists a huge gap between the understanding of a human being and a machine. Intelligent techniques bridge this gap by training machines to think and interpret a scene just like human beings do.

2. Intelligent techniques efficiently extract semantic features in an image which low-level feature-based techniques fail to extract.

3. Intelligent techniques involve making machines learn low-level features of image to construct feature vector which produces higher retrieval accuracy than the techniques involving processing of low-level features only.

CBIR using artificial intelligence techniques involves exploitation of techniques such as neural networks, genetic algorithms, artificial immune system, fuzzy logic, machine learning, and deep learning for constructing a feature vector. Lee and Yoo [48] proposed a CBIR technique using radial basis function network. This technique permits a user to initially select a query image and then use relevance feedback to incrementally search target images. Park et al. [49] proposed an image classification technique using neural network. The method first extracts shape-based texture features from wavelet transformed images and then constructs a neural network classifier for feature extraction using back propagation algorithm. Sadek et al. [50] proposed CBIR technique using cubic splines neural network. The method utilizes non-linear relationship between image features to compare similarity between images. Montazer and Giveki [51] proposed a CBIR technique by combining improved radial basis function networks and optimum steepest descent algorithm. Nagathan et al. [52] proposed feed-forward back propagation neural network approach for image retrieval.

Apart from neural networks, fuzzy logic is another intelligent technique which has been extensively used for CBIR. Vertan and Nezha [53] proposed the concept of fuzzy color histogram for retrieving similar images. Xiaoling and Kanglin [54] combined fuzzy logic with shape and color feature for CBIR. Chen and Wang [55] proposed a fuzzy logic approach named unified feature matching in which the image is represented by a set of segmented regions defined by a fuzzy feature set. Yager and Petry [56] combined relevance feedback and fuzzy logic to develop a framework for CBIR. Aboulmagd et al. [57] proposed a CBIR method using fuzzy logic which attempts to interpret an image similar to the way human beings interpret. This method represents image as Fuzzy Attributed Relational Graph (FARG), which computes attributes and spatial relation of each object in the image. The concept of FARG has also been used by Krishnapuram et al. [58] for constructing a feature vector for CBIR.

The trend of intelligence-based CBIR techniques also witnessed the use of genetic algorithm, artificial immune system, and machine learning techniques. Torres et al. [59] proposed a CBIR technique using genetic programming. The method exploits non-linear combination of image similarity to retrieve visually similar images. Analoui and Behesht [60] proposed an artificial immune system model for noisy image data sets. A number of techniques using support vector machine (SVM) have also been proposed for CBIR [61–64]. Such techniques utilize SVM to train the images of database and classify the query image in a relevant set using training data.

Intelligence-based techniques for CBIR are quite helpful in bridging the semantic gaps. The techniques discussed above tend to fulfill this objective efficiently and improve retrieval accuracy as compared to low-level feature-based techniques. However, a major drawback with these techniques is that they perform well on datasets of small size. They fail to perform efficiently on large datasets. These drawbacks are overcome by deep learning techniques. Deep learning techniques make use of deep architecture which learns feature at multiple level of abstractions [65]. Deep learning algorithms utilize deep architectures to perform modeling of high level of abstractions in data. This helps in bridging semantic gap more efficiently than traditional machine learning techniques. A number of image retrieval techniques based on deep learning models have been proposed in the recent years which attempt to demonstrate this fact through extensive experiments on large datasets. Lin et al. [66] proposed a deep learning framework in order to generate binary hash codes for faster image retrieval techniques. Saritha et al. [67] proposed a CBIR technique based on deep belief network for constructing feature vector. Zhao et al. [68] proposed deep semantic ranking-based technique in order to learn hash functions that are capable of preserving semantic similarity between multi-label images. Tzelepi and Tefas [69] proposed a CBIR technique that exploits deep CNN models for feature extraction. Fu et al. [70] combined the concept of deep convolutional network and support vector machine for image classification. Bai et al. [71] make use of deep convolutional neural network named AlexNet for extracting features from image for CBIR. Zhou et al. [72] utilized the concept of deep forest hashing for image retrieval. This method attempts to overcome the limitations of current deep neural network hashing models by using few hyperparameters and efficient training process.

The advantages of using deep learning techniques are that they perform well on large datasets and are capable of extracting features automatically instead of extracting in a handcrafted manner.

14.3.1 The Proposed Method

Intelligent techniques not only help in overcoming semantic gaps but also produce high retrieval accuracy as compared to low-level feature-based techniques. This section demonstrates the effectiveness of intelligent techniques in the field of CBIR. The method discussed in this section combines local binary pattern (LBP), HOG, and SVM to construct feature vector in order to retrieve visually similar images. HOG is a local shape descriptor, which efficiently extracts shape feature in an image without requiring segmentation of objects. Local binary pattern is an efficient descriptor which extracts texture feature in an image. The proposed method constructs feature vector by extracting shape feature from texture feature of image followed by training the feature vector of images with the help of SVM. Finally, SVM classifier is used to retrieve visually similar images using the trained feature vector of images in datasets. The proposed method consists of the following steps:

1. Computation of LBP codes of grayscale image.
2. Computation of HOG of resulting LBP codes.
3. Training of SVM by selecting samples from dataset of each classes.
4. Classification of images into relevant classes.

Following are the advantages of the proposed method:

1. It uses an HOG descriptor to extract shape feature which is an efficient local shape descriptor requiring no segmentation of objects unlike other shape descriptors such as moments.
2. It efficiently extracts texture feature through LBP which encodes relationship among pixel values in local neighborhood.
3. It uses SVM training to train and classify images in a dataset which help in bridging semantic gap.

14.3.2 Experiment and Results

In order to evaluate the effectiveness of the CBIR techniques, there are a number of image libraries containing wide variety of images such as Corel image library, Olivia image dataset, Caltech dataset, GHIM dataset, Coil-100 dataset, ImageNet, etc. To evaluate the performance of the proposed method, Corel-1K [75] image dataset has been used. Many researchers across the world are of the opinion that Corel dataset has the capability to meet all requirements for evaluating image retrieval systems due to its large size and diverse contents. The images in Corel-1K dataset are categorized into ten classes, namely Africans, Beaches, Buildings, Buses, Dinosaurs, Elephants, Flowers, Horses, Mountains, and Food. Each image is of size either 256×384 or 384×256 and each category consists of 100 images. To ease the experimentation process, the images in Corel-1K dataset were rescaled to size 256×256. The sample images from the dataset are shown in Figure 14.1. Each image of the dataset is taken as query image. If the retrieved images belong to the same category as that of the query image, the retrieval is considered to be successful. Otherwise, the retrieval is considered to have failed.

The performance of the proposed method has been measured in terms of precision and recall. Precision is defined as ratio of total number of relevant

FIGURE 14.1
Sample images from Corel-1K dataset.

images retrieved to the total number of relevant images in the dataset. Recall is defined as ratio of total number of relevant images retrieved to the total number of relevant images in the dataset.

14.3.3 Retrieval Result

LBP codes of gray scale images are computed followed by computation of HOG descriptors of resulting LBP codes. Finally, SVM is used to train the images of the dataset. The classification of images in relevant image class is performed using an SVM classifier.

Precision and recall have been computed for each category of image of Corel-1K dataset. Retrieval is considered to be good if the values of precision and recall are high. Table 14.1 shows the performance of the proposed method for each category of image of dataset in terms of precision and recall. Figure 14.2 shows plot between recall and category, and precision and category. The average values of recall for each category of image show what

TABLE 14.1

Average Precision and Recall for Each Category of Images of Corel-1K Dataset

S. No.	Category	Recall (%)	Precision (%)
1	Africans	29.92	36.22
2	Beaches	26.62	37.30
3	Buildings	33.64	44.40
4	Buses	36.00	45.70
5	Dinosaurs	36.57	57.20
6	Elephants	26.20	34.10
7	Flowers	40.75	55.20
8	Horses	27.31	35.90
9	Mountains	29.64	42.30
10	Food	25.77	36.40
Average		**31.24**	**42.66**

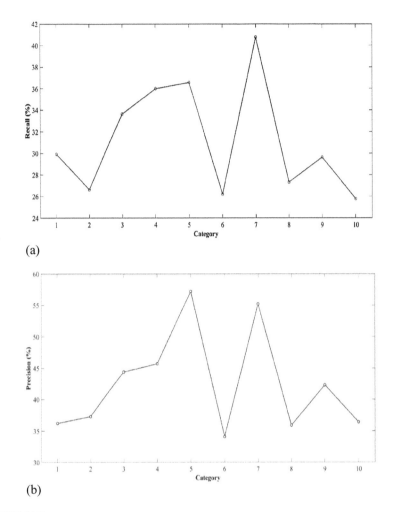

FIGURE 14.2
(a) Recall vs. category plot and (b) precision vs. category plot.

percentage of total number of relevant images in each category of image is being retrieved. The average values of recall for each category of image show what percentage of top ten images retrieved are relevant to the query image.

14.3.4 Performance Comparison

The proposed method attempts to incorporate intelligence-based technique to retrieve visually similar images. The effectiveness of the proposed method has been tested by comparing the retrieval accuracy of the proposed method with that of some of the other state-of-the-art CBIR methods which exploit low-level features to construct feature vector.

Table 14.2 and Figure 14.3 show the performance comparison of the proposed method with other state-of-the-art CBIR techniques. From Table 14.2 and Figure 14.3, it can be observed that the proposed method performs better than other CBIR techniques in terms of precision. The other CBIR techniques mostly use primary features only to construct feature vector. The low-level features fail to extract semantic aspect of image. The proposed technique incorporates intelligent method to retrieve visually similar images. It attempts to bridge the semantic gap by performing training and classification of images using the concept of SVM. Therefore, it produces high retrieval accuracy as compared to other CBIR techniques.

TABLE 14.2

Performance Comparison of the Proposed Method with Other State-of-the-Art CBIR Methods in Terms of Precision (%)

Method	Precision (%)
Takala et al. [73]	26.5
Yuan et al. [74]	23
Srivastava et al. [19]	35.94
Proposed method (without SVM)	26.85
Proposed method (with SVM)	**40.78**

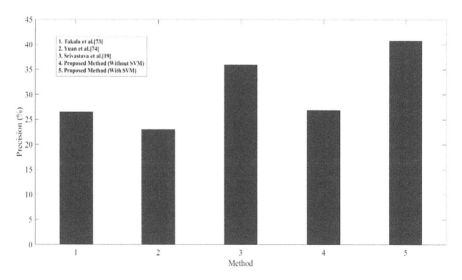

FIGURE 14.3
Performance Comparison of the Proposed Method with Other State-of-the-Art CBIR Methods in Terms of Precision (%).

14.4 Conclusion and Future Scope

This chapter deals with the concept of intelligent techniques for CBIR. The computational intelligence techniques are capable of capturing semantic features which are not efficiently captured by techniques based on low-level features. The effectiveness of intelligence-based techniques was demonstrated with the help of a proposed method which incorporates machine learning techniques to retrieve visually similar images. The proposed method extracts shape feature from texture feature using LBP. The classification of images in relevant classes has been performed using an SVM classifier. The proposed method attempts to bridge the semantic gap as it incorporates multiple features of image along with intelligent technique. Therefore, it produces high retrieval accuracy when compared with other low-level feature-based techniques. There can be further improvement in the proposed method by incorporating multiresolution techniques to extract semantic features at more than one resolution of image.

Computational intelligence techniques not only find applications in the field of visual information retrieval but in other fields of computer vision also. While low-level features efficiently determine visual characteristics of an image, the high-level features are quite efficient in representing semantic aspect of image. Much of the research has been done to propose techniques which exploit low-level features of image, but feature extraction using semantic features is still an open problem of research. The CBIR techniques in the past have mostly used traditional machine learning techniques which have been successful in overcoming semantic gaps on small image datasets. However, the recent techniques of CBIR have witnessed the use of advanced intelligent techniques such as deep learning. These techniques have been proved to be quite efficient on very large datasets containing large number of images. These techniques can prove to be quite effective in overcoming semantic gaps. Automation is going to be the future of industries and information retrieval will play a key role in carrying out important industrial operations. In order to carry out operations in an automated way which, at present, are executed with human intervention, there is a need to overcome semantic gap. Modern computational intelligence techniques can prove to be quite helpful in bridging the semantic gaps.

References

[1] Smeulders, Arnold WM, Marcel Worring, Simone Santini, Amarnath Gupta, and Ramesh Jain. Content-based image retrieval at the end of the early years. *IEEE Transactions on Pattern Analysis and Machine Intelligence*, 22(12) (2000): 1349–1380.

[2] Lin, Chuen-Horng, Rong-Tai Chen, and Yung-Kuan Chan. A smart content-based image retrieval system based on color and texture feature. *Image and Vision Computing*, 27(6) (2009): 658–665.

[3] Osadebey, M. E. (2006). *Integrated Content-based Image Retrieval using Texture, Shape and Spatial Information*. Umea University, Umea.

[4] Hiremath, P. S., and Jagadeesh Pujari. *Content based image retrieval using color, texture and shape features*. In *15th International Conference on Advanced Computing and Communications (ADCOM 2007)*, pp. 780–784. IEEE, 2007.

[5] Lamard, Mathieu, Guy Cazuguel, Gwénolé Quellec, Lynda Bekri, Christian Roux, and Béatrice Cochener. *Content based image retrieval based on wavelet transform coefficients distribution*. In *2007 29th Annual International Conference of the IEEE Engineering in Medicine and Biology Society*, pp. 4532–4535. IEEE, 2007.

[6] Quellec, Gwénolé, Mathieu Lamard, Guy Cazuguel, Béatrice Cochener, and Christian Roux. Adaptive nonseparable wavelet transform via lifting and its application to content-based image retrieval. *IEEE Transactions on Image Processing*, 19(1) (2009): 25–35.

[7] Loupias, Etienne, Nicu Sebe, Stéphane Bres, and J.-M. Jolion. *Wavelet-based salient points for image retrieval*. In *Proceedings 2000 International Conference on Image Processing (Cat. No. 00CH37101)*, 2, pp. 518–521. IEEE, 2008.

[8] Fadaei, Sadegh, Rassoul Amirfattahi, and Mohammad Reza Ahmadzadeh. New content-based image retrieval system based on optimised integration of DCD, wavelet and curvelet features. *IET Image Processing*, 11(2) (2016): 89–98.

[9] Alzu'bi, Ahmad, Abbes Amira, and Naeem Ramzan. Semantic content-based image retrieval: A comprehensive study. *Journal of Visual Communication and Image Representation*, 32 (2015): 20–54.

[10] Sergyan, Szabolcs. *Color histogram features based image classification in content-based image retrieval systems*. In *2008 6th International Symposium on Applied Machine Intelligence and Informatics*, pp. 221–224. IEEE, 2008.

[11] Liu, Guang-Hai, and Jing-Yu Yang. Content-based image retrieval using color difference histogram. *Pattern Recognition*, 46(1) (2013): 188–198.

[12] Huang, Jing, S. Ravi Kumar, and Mandar Mitra. *Combining supervised learning with color correlograms for content-based image retrieval*. In *Proceedings of the fifth ACM International Conference on Multimedia*, pp. 325–334, 1997.

[13] Pass, Greg, Ramin Zabih, and Justin Miller. *Comparing images using color coherence vectors*. In *Proceedings of the fourth ACM International Conference on Multimedia*, pp. 65–73, 1997.

[14] Ojala, Timo, Matti Pietikäinen, and Topi Mäenpää. *Gray scale and rotation invariant texture classification with local binary patterns*. In *European Conference on Computer Vision*, pp. 404–420. Springer, Berlin, Heidelberg, 2000.

[15] Tan, Xiaoyang, and Bill Triggs. Enhanced local texture feature sets for face recognition under difficult lighting conditions. *IEEE Transactions on Image Processing*, 19(6) (2010): 1635–1650.

[16] Zhang, Baochang, Yongsheng Gao, Sanqiang Zhao, and Jianzhuang Liu. Local derivative pattern versus local binary pattern: face recognition with high-order local pattern descriptor. *IEEE Transactions on Image Processing*, 19(2) (2009): 533–544.

[17] Murala, Subrahmanyam, R. P. Maheshwari, and R. Balasubramanian. Local tetra patterns: a new feature descriptor for content-based image retrieval. *IEEE Transactions on Image Processing*, 21(5) (2012): 2874–2886.

[18] Yap, P.-T., and R. Paramesran. Content-based image retrieval using Legendre chromaticity distribution moments. *IEE Proceedings-Vision, Image and Signal Processing*, 153(1) (2006): 17–24.

[19] Srivastava, Prashant, Nguyen Thanh Binh, and Ashish Khare. *Content-based image retrieval using moments*. In *International Conference on Context-Aware Systems and Applications*, pp. 228–237. Springer, Cham, 2013.

[20] Dalal, Navneet, and Bill Triggs. *Histograms of oriented gradients for human detection*. In *2005 IEEE Computer Society Conference on Computer Vision and Pattern Recognition (CVPR'05)*, vol. 1, pp. 886–893. IEEE, 2005.

[21] Junior, Oswaldo Ludwig, David Delgado, Valter Gonçalves, and Urbano Nunes. *Trainable classifier-fusion schemes: An application to pedestrian detection*. In *2009 12th international IEEE Conference on Intelligent Transportation Systems*, pp. 1–6. IEEE, 2009.

[22] Andreou, Ioannis, and Nikitas M. Sgouros. Computing, explaining and visualizing shape similarity in content-based image retrieval. *Information Processing & Management*, 41(5) (2005): 1121–1139.

[23] Yue, Jun, Zhenbo Li, Lu Liu, and Zetian Fu. Content-based image retrieval using color and texture fused features. *Mathematical and Computer Modelling*, 54(3–4) (2011): 1121–1127.

[24] Chun, Young Deok, Nam Chul Kim, and Ick Hoon Jang. Content-based image retrieval using multiresolution color and texture features. *IEEE Transactions on Multimedia*, 10(6) (2008): 1073–1084.

[25] Fu, Xuezheng, Yong Li, Robert Harrison, and Saeid Belkasim. *Content-based image retrieval using gabor-zernike features*. In *18th International Conference on Pattern Recognition (ICPR'06)*, 2, pp. 417–420. IEEE, 2006.

[26] Srivastava, Prashant, Nguyen Thanh Binh, and Ashish Khare. Content-based image retrieval using moments of local ternary pattern. *Mobile Networks and Applications*, 19, 5 (2014): 618–625.

[27] Alsmadi, Mutasem K. Content-Based Image Retrieval Using Color, Shape and Texture Descriptors and Features. *Arabian Journal for Science and Engineering*, (2020): 1–14.

[28] Nazir, Atif, Rehan Ashraf, Talha Hamdani, and Nouman Ali. *Content based image retrieval system by using HSV color histogram, discrete wavelet transform and edge histogram descriptor*. In *2018 International Conference on Computing, Mathematics and Engineering Technologies (iCoMET)*, pp. 1–6. IEEE, 2011.

[29] Yildizer, Ela, Ali Metin Balci, Tamer N. Jarada, and Reda Alhajj. Integrating wavelets with clustering and indexing for effective content-based image retrieval. *Knowledge-Based Systems*, 31 (2012): 55–66.

[30] Srivastava, Prashant, and Ashish Khare. Integration of wavelet transform, local binary patterns and moments for content-based image retrieval. *Journal of Visual Communication and Image Representation*, 42 (2017): 78–103.

[31] Singha, Manimala, K. Hemachandran, and A. Paul. Content-based image retrieval using the combination of the fast wavelet transformation and the colour histogram. *IET Image Processing*, 6(9) (2012): 1221–1226.

[32] Srivastava, Prashant, and Ashish Khare. Content-based image retrieval using local ternary wavelet gradient pattern. *Multimedia Tools and Applications*, 78(24) (2019): 34297–34322.

[33] Xavier, Lidiya, Bella Mary I. Thusnavis, and David Raj W. Newton. *Content based image retrieval using textural features based on pyramid-structure wavelet transform*. In *2011 3rd International Conference on Electronics Computer Technology*, 4, pp. 79–83. IEEE, 2011.

[34] Khare, Manish, Prashant Srivastava, Jeonghwan Gwak, and Ashish Khare. *A multiresolution approach for content-based image retrieval using wavelet transform of local binary pattern*. In *Asian Conference on Intelligent Information and Database Systems*, pp. 529–538. Springer, Cham, 2018.

[35] Moghaddam, Hamid Abrishami, Taher Taghizadeh Khajoie, and Amir Hossein Rouhi. A new algorithm for image indexing and retrieval using wavelet correlogram. In *Proceedings 2003 International Conference on Image Processing* (Cat. No. 03CH37429), 3, pp. 3–497. IEEE, 2003.

[36] Srivastava, Prashant, Om Prakash, and Ashish Khare. *Content-based image retrieval using moments of wavelet transform*. In *The 2014 International Conference on Control, Automation and Information Sciences (ICCAIS 2014)*, pp. 159–164. IEEE, 2014.

[37] Srivastava, Prashant, Manish Khare, and Ashish Khare. *Combining Local Binary Pattern and Speeded-Up Robust Feature for Content-Based Image Retrieval*. In *Asian Conference on Intelligent Information and Database Systems*, pp. 366–376. Springer, Singapore, 2020.

[38] Sumana, Ishrat Jahan, Md Monirul Islam, Dengsheng Zhang, and Guojun Lu. *Content based image retrieval using curvelet transform*. In *2008 IEEE 10th Workshop on Multimedia Signal Processing*, pp. 11–16. IEEE, 2008.

[39] Youssef, Sherin M. ICTEDCT-CBIR: Integrating curvelet transform with enhanced dominant colors extraction and texture analysis for efficient content-based image retrieval. *Computers & Electrical Engineering*, 38(5) (2012): 1358–1376.

[40] Srivastava, Prashant, and Ashish Khare. Content-based image retrieval using local binary curvelet co-occurrence pattern—a multiresolution technique. *The Computer Journal*, 61(3) (2018): 369–385.

[41] Sumana, Ishrat Jahan, Guojun Lu, and Dengsheng Zhang. *Comparison of curvelet and wavelet texture features for content based image retrieval*. In *2012 IEEE International Conference on Multimedia and Expo*, pp. 290–295. IEEE, 2012.

[42] Gonde, Anil Balaji, R. P. Maheshwari, and R. Balasubramanian. Modified curvelet transform with vocabulary tree for content based image retrieval. *Digital Signal Processing*, 23(1) (2013): 142–150.

[43] Zhang, Dengsheng, M. Monirul Islam, Guojun Lu, and Ishrat Jahan Sumana. Rotation invariant curvelet features for region based image retrieval. *International Journal of Computer Vision*, 98(2) (2012): 187–201.

[44] Arun, K. S., and Hema P. Menon. Content based medical image retrieval by combining rotation invariant contourlet features and fourier descriptors. *International Journal of Recent Trends in Engineering*, 2(2) (2009): 35.

[45] Romdhane, Rim, Hela Mahersia, and Kamel Hamrouni. *A novel content image retrieval method based on contourlet*. In *2008 3rd International Conference on Information and Communication Technologies: From Theory to Applications*, pp. 1–5. IEEE, 2008.

[46] Liu, Ying, Dengsheng Zhang, Guojun Lu, and Wei-Ying Ma. A survey of content-based image retrieval with high-level semantics. *Pattern Recognition*, 40 (2007): 262–282.

[47] Eakins, John P. Towards intelligent image retrieval. *Pattern Recognition*, 35(1) (2002): 3–14.

[48] Lee, Hyoung Ku, and Suk In Yoo. Intelligent image retrieval using neural network. *IEICE Transactions on Information and Systems*, 84(12) (2001): 1810–1819.

[49] Park, Soo Beom, Jae Won Lee, and Sang Kyoon Kim. Content-based image classification using a neural network. *Pattern Recognition Letters*, 25(3) (2004): 287–300.

[50] Sadek, Samy, Ayoub Al-Hamadi, Bernd Michaelis, and Usama Sayed. Image retrieval using cubic splines neural networks. *International Journal of Video & Image Processing and Network Security (IJIPNS)*, 9(10) (2009): 17–22.

[51] Montazer, Gholam Ali, and Davar Giveki. An improved radial basis function neural network for object image retrieval. *Neurocomputing*, 168(2015): 221–233.

[52] Nagathan, Arvind, and Jitendranath Mungara. Content-based image retrieval system using feed-forward backpropagation neural network. *International Journal of Computer Science and Network Security (IJCSNS)*, 14(6) (2014): 70.

[53] Vertan, Constantin, and Nozha Boujemaa. *Embedding fuzzy logic in content based image retrieval*. In *PeachFuzz 2000. 19th International Conference of the North American Fuzzy Information Processing Society-NAFIPS* (Cat. No. 00TH8500), pp. 85–89. IEEE, 2000.

[54] Xiaoling, Wang, and Xie Kanglin. Application of the fuzzy logic in content-based image retrieval. *Journal of Computer Science & Technology*, 5 (2005).

[55] Chen, Yixin, and James Ze Wang. A region-based fuzzy feature matching approach to content-based image retrieval. *IEEE Transactions on Pattern Analysis and Machine Intelligence*, 24(9) (2002): 1252–1267.

[56] Yager, Ronald R., and Frederick E. Petry. A framework for linguistic relevance feedback in content-based image retrieval using fuzzy logic. *Information Sciences*, 173(4) (2005): 337–352.

[57] Aboulmagd, Heba, Neamat El-Gayar, and Hoda Onsi. A new approach in content-based image retrieval using fuzzy. *Telecommunication Systems*, 40(1–2) (2009): 55.

[58] Krishnapuram, Raghu, Swarup Medasani, Sung-Hwan Jung, Young-Sik Choi, and Rajesh Balasubramaniam. Content-based image retrieval based on a fuzzy approach. *IEEE Transactions on Knowledge and Data Engineering* 16(10) (2004): 1185–1199.

[59] Torres, Ricardo da S., Alexandre X. Falcão, Marcos A. Gonçalves, João P. Papa, Baoping Zhang, Weiguo Fan, and Edward A. Fox. A genetic programming framework for content-based image retrieval. *Pattern Recognition*, 42(2) (2009): 283–292.

[60] Analoui, Morteza, and Maedeh Beheshti. *Content-based image retrieval using artificial immune system(ais) clustering algorithms*. In *World Congress on Engineering 2012*, July 4–6, 2012. London, 2188, pp. 388–393. International Association of Engineers, 2010.

[61] Rao, Ch, S. Srinivas Kumar, and B. Chandra Mohan. Content based image retrieval using exact legendre moments and support vector machine. arXiv preprint arXiv:1005.5437 (2010).

[62] Sharif, Uzma, Zahid Mehmood, Toqeer Mahmood, Muhammad Arshad Javid, Amjad Rehman, and Tanzila Saba. Scene analysis and search using local features and support vector machine for effective content-based image retrieval. *Artificial Intelligence Review* 52(2) (2019): 901–925.

[63] Seo, Kwang-Kyu. An application of one-class support vector machines in content-based image retrieval. *Expert Systems with Applications*, 33(2) (2007): 491–498.

[64] Zhang, Lei, Fuzong Lin, and Bo Zhang. Support vector machine based relevance feedback algorithm in image retrieval. *Journal-Tsinghua University*, 42, 1 (2002): 80–83.

[65] Wan, Ji, Dayong Wang, Steven Chu Hong Hoi, Pengcheng Wu, Jianke Zhu, Yongdong Zhang, and Jintao Li. *Deep learning for content-based image retrieval: A comprehensive study*. In *Proceedings of the 22nd ACM international conference on Multimedia*, pp. 157–166. 2014.

[66] Lin, Kevin, Huei-Fang Yang, Jen-Hao Hsiao, and Chu-Song Chen. *Deep learning of binary hash codes for fast image retrieval*. In *Proceedings of the IEEE conference on computer vision and pattern recognition workshops*, pp. 27–35. 2015.

[67] Saritha, R. Rani, Varghese Paul, and P. Ganesh Kumar. Content based image retrieval using deep learning process. *Cluster Computing*, 22, 2 (2019): 4187–4200.

[68] Zhao, Fang, Yongzhen Huang, Liang Wang, and Tieniu Tan. *Deep semantic ranking based hashing for multi-label image retrieval*. In *Proceedings of the IEEE Conference on Computer Vision and Pattern Recognition*, pp. 1556–1564. 2015.

[69] Tzelepi, Maria, and Anastasios Tefas. Deep convolutional learning for content based image retrieval. *Neurocomputing*, 275 (2018): 2467–2478.

[70] Fu, Ruigang, Biao Li, Yinghui Gao, and Ping Wang. *Content-based image retrieval based on CNN and SVM*. In *2016 2nd IEEE International Conference on Computer and Communications (ICCC)*, pp. 638–642. IEEE, 2016.

[71] Bai, Cong, Ling Huang, Xiang Pan, Jianwei Zheng, and Shengyong Chen. Optimization of deep convolutional neural network for large scale image retrieval. *Neurocomputing*, 303 (2018): 60–67.

[72] Zhou, Meng, Xianhua Zeng, and Aozhu Chen. Deep forest hashing for image retrieval. *Pattern Recognition*, 95 (2019): 114–127.

[73] Takala, Valtteri, Timo Ahonen, and Matti Pietikäinen. *Block-based methods for image retrieval using local binary patterns*. In *Scandinavian Conference on Image Analysis*, pp. 882–891. Springer, Berlin, Heidelberg, 2005.

[74] Yuan, Xiaoli, Jing Yu, Zengchang Qin, and Tao Wan. *A SIFT-LBP image retrieval model based on bag of features*. In *IEEE International Conference on Image Processing*, pp. 1061–1064, 2011.

[75] http://wang.ist.psu.edu/docs/related/ (Last accessed on 7.10.2020).

Index

A

accuracy, 40, 96, 100–101, 106–111
age-related macular degeneration
    (AMD), 94–95
aggregating rule strength, 231
algorithm, 76, 78, 80–83, 85, 89
ANN, *see* artificial neural netwok
artificial intelligence (AI), 40, 42
artificial neural netwok (ANN), 58, 79,
    82, 86–88
automatic emotion detection, 48
autonomous, 37, 43
average link clustering, 177

B

Bagging Regressor, 206–207, 209,
    214–216
Bayesian optimization, 205, 208, 210–211
better human-computer interaction
    systems, 220
Big Data, 159
blindness, 94, 98
bounding boxes, 40–41

C

Catboost, 206, 209, 214–216
classification, 77, 81, 83, 88, 93, 96–97,
    99–102, 104–106, 108, 110
clustering, 177
CNN, *see* convolutional neural network
cold start, 161
collaborative filtering, 161
computational intelligence, 48, 50, 52–53
content-based filtering, 161
content-based image retrieval, 257–258
Contrast Enhancement Dynamic
    Histogram Equalization,
    63–64, 67

convolutional layers, 37, 42
convolutional neural network (CNN),
    38, 40–41, 44, 77–82, 86, 96,
    103–104, 107–111
corpus filtering and preprocessing, 227
cosine similarity, 172
cross entropy loss, 42
cryptography, 124

D

Data as a Service, 118
Data Encryption Algorithm, 127
data privacy, 124
deep learning, 38, 40, 79, 82, 85–86, 89,
    93, 103–104, 107, 111, 263
defuzzification, 231
demographic filtering, 161
diabetes mellitus (DM), 94–99, 111
diabetic mellitus retinopathy (DMR), 94
diabetic retinopathy (DR), 94–104, 110
dynamic histogram equalization, 63–67

E

EEG, 48–53
EfficientNet, 96, 104–105, 107–111
embedding, 17, 20–21, 27
emotion, 48, 51–53, 58–59, 76–82, 84, 86,
    88–89
ensembles, 206, 208
epoch, 42
Extra Tree Regressor, 215–216
EyePACS, 107, 109–110

F

F1 score, 42, 44
facial detection, 76–77, 80, 85
facial gestures, 75, 79–80
fatality rate, 38

feature engineering, 201
feature extraction, 51–52, 55, 63–65, 67,
        69, 71
finger vein, 63–69, 71
firebase, 43
fully homomorphic encryption, 129
fundus, 94, 96, 98–99, 103–104, 109–110

G

galvanic skin response(GSR), 78, 81, 89
Gaussian noise, 40
gestures, 75–76, 79–82, 86, 88–90
glaucoma, 94, 100–101, 111
GPS, 38, 43
grayscale, 40

H

Hadoop, 119, 162
homomorphic encryption algorithm, 125
hybrid systems approach, 222
hyperparameter tuning, 199, 206, 208

I

ICA, 50, 53
identification, 63–64, 71
image encryption, 16
image enhancement, 63–65, 67, 69, 71
imageNet, 41
image processing, 16, 18, 21, 58, 77–78, 88
image retrieval, 258
image security, 21
Infrastructure as a Service, 118
infrastructure security, 123
intelligent techniques, 259, 261–263

K

key rotation algorithm, 127
keyword matching, 171
KNN, 78
Kohonen networks, 179

L

learning rate, 42
lemmatization, 224

LightGBM, 198, 205, 209, 212, 214–216
Linear Regressor, 214–216
linguistic resources, 225
logistic regression, 222
long short-term memory (LSTM), 77, 81,
        88–89
LVO, 55

M

machine learning, 78, 80–85, 89, 94, 96, 99, 108
MapReduce, 119, 162
mean square error, 63, 69
message digest algorithm, 126
mood, 75–79, 81–83, 86, 88–90
mosaic image, 17, 31
multilayer perceptron, 77, 87
multiresolution techniques, 259–260

N

naïve Bayes, 222
natural language processing, 222
nearest neighbour, 172
network attached storage, 116
neural network, 77–82, 85–88

O

object detection, 38
ophthalmologist, 94, 97, 105–106
opinion analysis, 220
output membership function, 231

P

partially homomorphic encryption, 129
Peak Signal to Noise Ratio, 63, 69
Platform as a Service, 117
playlist, 76–81, 90
polarity, 220
precision, 106, 109–110, 264–265, 267
profile exploitation, 169
proliferative DR, 96, 98, 100–102, 107

R

random forest, 78, 82, 85
real time object detection, 37

recall, 264–266
recognition, 64, 67, 71
recurrent neural network (RNN), 40
referable diabetic retinopathy (RDR), 107–108
RepeatedKFold, 208–209
ResNet, 38, 41
retinal image quality assessment (RIQA), 97
retinopathy, 94–95, 97–100, 102–103, 107
Rijndael encryption algorithm, 127
Rivest–Shamir–Adleman, 125
root mean square error, 181

S

semantic relationship, 222
sensitivity, 100, 106, 110
sentiment analysis using fuzzy logic, 229
SentiWordNet, 226
severity, 96, 98, 101, 103, 111
SoftMax, 42
Software as a Service, 117
song recommendation, 89–90
specificity, 105–106, 110
spectral clustering, 178
speech recognition, 76, 78
stacking regressor, 214
state of the art, 44
steganography, 16–18, 20–21
supervised, 83–84

support vector machine (SVM), 77–78, 81, 83–84, 96, 99–101, 222, 263–265, 267–268

T

text to speech generation, 220
threshold, 38
transfer learning, 38
true negative (TN), 106
true positive (TP), 106

V

VADER, 226
verification, 64

W

WMNE algorithm, 54
WordNet, 225
word study, 228

X

XGBoost, 198, 203, 205, 209, 213–216

Y

YOLO, 37, 40, 44

For Product Safety Concerns and Information please contact our EU
representative GPSR@taylorandfrancis.com
Taylor & Francis Verlag GmbH, Kaufingerstraße 24, 80331 München, Germany

www.ingramcontent.com/pod-product-compliance
Ingram Content Group UK Ltd.
Pitfield, Milton Keynes, MK11 3LW, UK
UKHW051942210425
457613UK00026BA/129